Markets and States
in Tropical Africa

**California Series on
Social Choice and Political Economy**

edited by Brian Barry and Samuel L. Popkin

Markets and States in Tropical Africa

The Political Basis of Agricultural Policies

Robert H. Bates

UNIVERSITY OF CALIFORNIA PRESS

Berkeley • Los Angeles • London

University of California Press
Berkeley and Los Angeles, California

University of California Press, Ltd.
London, England

© 1981 by
The Regents of the University of California

First Paperback Printing 1984
ISBN 0-520-05229-3

Printed in the United States of America

1   2   3   4   5   6   7   8   9

Library of Congress Cataloging in Publication Data

Bates, Robert H
    Markets and states in tropical Africa.
    (California series on social choice and political
economy)

    Bibliography: p.
    Includes index.
    1. Agriculture and state—Africa, Sub-Saharan.
    2. Agriculture-Economic aspects—Africa, Sub-Saharan.
    I. Title.   II. Series
    HD2118.1981.B37   338.1'867   80-39732

To Margaret

Contents

Acknowledgments

My research has received the assistance of numerous persons and organizations, both in Africa and the United States.

I wish to acknowledge the assistance I received in Ghana from the Department of Agricultural Economy and Farm Management of the University of Ghana; the Institute for Statistical, Social, and Economic Research; the Ministry of Agriculture; the Agricultural Development Bank; the United States Agency for International Development; the International Bank for Reconstruction and Development; and the Ministry of Cocoa Affairs. I particularly wish to thank Michael and Mary Warren and Professor Emmanuel Andah for their aid and hospitality.

I am deeply grateful for the help I received in Nigeria from the Rockefeller Foundation, the Institute of Tropical Africa, and the University of Ibadan, particularly the Departments of Politics and Agricultural Economics and the Institute of Social and Economic Research. I give particular thanks to Professor Tyler Biggs for assistance during my stay. The generous advice of Elon Gilbert and Samson Olayidi were critical to the success of my work in Nigeria.

For help during my work in Kenya, I wish to thank the Ford Foundation, the Institute for Development Studies, the Departments of Law and Political Science at the University of Nairobi, the planning unit of the Ministry of Agriculture, and the libraries of the Ministry of Agriculture, the Central Bureau of Statistics, the Institute for Development Studies, and the National Archives. My special thanks go to Suzanne Drouilh for her hospitality and guidance. I wish also to note the help of Raphael Kaplinsky, Timothy Aldington, Edgar Winnans, and Jennifer Sharpley. David Brokensha was instrumental in assisting my work in Kenya.

In Tanzania, I received generous assistance from the Marketing Research Bureau of the Ministry of Agriculture. I wish in particular to thank Stephen Lombard for his aid and encouragement.

I wish also to acknowledge the assistance I received from the International Coffee Organization and the International Cocoa Organization in London.

I became committed to this subject while a visiting scholar at the Food Research Institute of Stanford University. I wish to thank Walter Falcon, the Director; the faculty and staff of the Institute; and, in particular, the African specialists, William Jones, Bruce Johnston, and Scott Pearson. I also wish to thank the Social Science Research Council and the California Institute of Technology for supporting my work at Stanford.

My interest in the subject was reinforced during the spring of 1978 when Michael Lofchie and I offered a seminar on African agricultural development at the African Studies Center of the University of California, Los Angeles. I am most grateful to the participants in that seminar, and especially to Michael Lofchie, for the generosity with which they shared their knowledge, criticism, and encouragement.

I have received support for this work from the Division of Humanities and Social Sciences of the California Institute of Technology. I wish in particular to thank Marcia Nelson and Jo Azary for preparing the manuscript; the Munger Africana Library; and especially the expert and cheerful assistance of the Interlibrary Loan Department of Millikan Library. I am grateful as well for the research assistance of Leslie Madden, Mark Granger, and especially

Miriam Eichwald. William Rogerson and Kenneth McCue have stimulated and sharpened my thinking at numerous points in my argument.

I received critical input for this work from participants in seminars at the African Studies Center of the University of California, Los Angeles; the Institute of International Studies, University of California, Berkeley; the 1979 Annual Meeting of the American Political Science Association; and the Interdisciplinary Institute of Urban and Regional Studies of the Economic University of Vienna.

The manuscript has benefited immeasurably from comments and criticisms provided by David Abernethy, Barry Ames, Gordon Appleby, Brian Barry, Bruce Cain, Carl Eicher, John Ferejohn, Morris Fiorina, Elon Gilbert, Yujiro Hayami, Frances Hill, Goran Hyden, Bruce Johnston, William Jones, Marvin Miracle, Gary Miller, Roger Noll, Joe Oppenheimer, Samuel Popkin, James Scott, Thayer Scudder, Alan Sweezy, Judith Tendler, Gordon Tullock, and two anonymous readers from the University of California Press. I owe special thanks to Samuel Popkin for his encouragement and criticism.

Lastly, I wish to thank the National Science Foundation for its support, as conferred through Grant No. SOC 77-08573A1.

Introduction

Over the last decade a deepening sense of crisis has arisen among those concerned with African agriculture. The Sahelian drought of the mid-1970s temporarily dramatized the plight of the African farmer, but those who follow Africa are convinced that the problems lie deeper than the vagaries of the weather. They cite figures published by the Food and Agricultural Organization of the United Nations, which suggest that per capita food production in Africa had been stagnating prior to the drought; and they stress that although the rains have returned, food production has failed to recover (FAO 1978b, pp. 77–78). They also note that in a continent peopled largely by farmers, an ever-increasing portion of scarce foreign exchange is being spent on imports of food.

It is also apparent that Africa's agricultural exports have decreased. Palm oil in Nigeria, groundnuts in Senegal, cotton in Uganda, and cocoa in Ghana were once among the most prosperous industries in Africa. But in recent years, farmers of these crops have produced less, exported less, and earned less in foreign markets. What is true in these cases appears to be true elsewhere: the studies of the Food and Agricultural Organization indicate that during the 1970s the volume of agricultural exports from all of Africa has declined. There are, of course, many reasons for the apparent shortfalls in agricultural production in Africa. This book addresses the political origins of the problem.

THE SEARCH FOR A PARADIGM

The need for increased farm production is not unique to Africa, nor are the factors inhibiting agricultural development confined to that continent alone. For decades, researchers have examined difficulties bedeviling the growth of farming in Third World nations. They have pointed to factors originating in the physical and biological environments of farmers, and they have isolated social and economic impediments as well. A consensus has emerged that the most important of these factors is the nature of the incentives offered to producers. Physical and biological factors, it is held, are constraints that farmers can transcend, provided they are given sufficient incentive to do so. In the words of Theodore Schultz, perhaps the most noted proponent of this argument: "Incentives to guide and reward farmers are a critical component. Once there are investment opportunities and efficient incentives, farmers will turn sand into gold" (1976, p. 5).

If the basic problem of farming in the developing countries is improper incentives for farmers, then it follows that the origins of the problem lie in the actions of those who distort the operations of the market. Schultz and his followers have been quick to identify one major source of such distortions: the policies adopted by governments. By adopting policies that confound the operation of markets, Third World governments undercut the productive potential of their farm populations. A major source of the problem of Third World agriculture is bad policy (Schultz 1978).

But agricultural economists who stress the ineptness of government policy-makers often praise the acumen of persons in market operations. To put it bluntly, people are said to display both economic shrewdness and political stupidity. There is thus a major inconsistency; and because those who advance these arguments fail to identify the source of this puzzling imbalance in human capabilities, their analysis remains incomplete.

This work seeks to go beyond the position of the agricultural economists by asking the obvious question: Why should reasonable men adopt public policies that have harmful consequences for the societies they govern? In answering this question, it looks for the social purposes that lead policy-makers to intervene in agricultural markets. Above all, it examines the political calculations that induce governments to intervene in ways which are harmful to the interests of most farmers.

METHOD AND SCOPE

The starting point of our analysis is a mental artifact—a conception of the economic location of agricultural producers. Farmers are seen as standing at the intersection of three major markets. Their real incomes depend upon their performance in these markets. They derive their revenues from the sales they make in the first of these markets—the market for agricultural commodities. Their profits are a function of these revenues, but also of the costs they incur in a second major market—the market for factors of production. And the real value of their profits, and thus the real value of their incomes, is determined by the prices they must pay in a third major market: the market for consumer goods, particularly commodities manufactured in the city.

Our conceptualization of the agrarian producer suggests a definition of agricultural policy. Agricultural policy consists of governmental actions that affect the incomes of rural producers by influencing the prices these producers confront in the major markets which determine their incomes. In describing the agricultural policies of African states, we therefore examine government intervention in three markets: the markets for agricultural commodities, the markets for inputs into farming, and the markets for the goods that farmers buy from the urban-industrial sector.

One major purpose of this book is to account for the policies adopted by African governments. We assume throughout that political action is purposeful behavior, and that among the major purposes of governments are the pursuit of certain social objectives and the resources needed to achieve them. Foremost among the social objectives of governments in the developing areas is to shift the basis of their economies away from the production of agricultural commodities and toward the production of manufactured goods. This objective strongly influences their choice of agricultural policies.

This is as true of governments in Africa as it is of other governments in the developing areas. They intend to transform their economies; they want to move resources from agriculture to industry; and therefore they set prices in markets in order to capture resources from agriculture. Moreover, the governments need resources with which to implement these development programs; and to achieve their objectives, they need foreign exchange. In nations in which agriculture is the greatest source of income and the principal source of exports, it is natural that they should seek to levy revenues from the rural sector. Out of a commitment to development, governments in Africa therefore intervene in agricultural markets and extract the resources they need to build a "modern" economy.

While acknowledging the importance of public purposes and reasons of state in motivating agricultural policy, we also recognize that more personal motives animate political choices. Governments want to stay in power. They must appease powerful interests. And people turn to political action to secure special advantages—rewards they are unable to secure by competing in the marketplace. This book stresses the role of such factors in the formulation of agricultural policy.

It is critical that analysts recognize the importance of these forces. One obvious conclusion sometimes drawn from the economists' critique of Third-World agricultural programs, for example, is that governments should withdraw from agricultural markets and let economic forces prevail. Such counsel is naturally ignored by policy-makers as hopelessly naive. Similarly, although governments intervene in markets to secure social objectives, it would be unrealistic to believe that these public objectives are the sole force behind their choices. For to secure any given objective, governments can choose from a variety of techniques. Let us glance at some examples.

To increase food supplies, governments could offer higher prices for food, or they could invest the same amount of resources in food production projects. There is every reason to believe that pricing policies are the more efficient way of securing the objective. But governments in Africa systematically prefer project-based policies to price-based policies. I shall argue that they do so because they find project-based policies politically more useful.

To strengthen the incentives for food production, governments can increase the prices of farm products, or they can subsidize the costs of farm implements. Either action would result in higher profits for producers, but governments prefer the latter policy. They do so in part because of its superior political attractions.

Agriculture in Africa is both subsidized and taxed. Governments tax the products of farmers while subsidizing the inputs they employ. This apparently paradoxical mixture becomes reasonable when viewed from a political perspective.

To increase output, governments finance production programs. But in doing so, they introduce characteristic distortions. Given the level of resources devoted to the programs, for example, they often create too many projects; the programs then fail because the resources have been spread too thin. Such behavior is nonsensical when analyzed solely in terms of stated objectives, but it becomes understandable once we consider the political calculations underlying the choices of governments.

Consider a final example. In the face of shortages, governments can allow prices to rise, or they can maintain lower prices while imposing quotas. In a variety of markets that are of significance to agricultural producers, African governments choose to ration; and when they do so, they give no systematic preference to the poor. Their use of nonmarket mechanisms in the face of shortages reflects not their social values but their calculations of how their political interests can best be served.

These examples suggest the importance of examining the political basis for the selection of agricultural policies. This book examines the content of agricultural policies in Africa. And in doing so, it stresses the ways in which policies are designed to secure advantages for particular interests, to appease powerful political forces, and to enhance the capacity of political regimes to remain in power.

THEMES AND BROADER SETTING

Four other main themes will appear in this book. The first concerns the feelings of disillusion now prevalent in much of Africa. Politically sensitive persons discern the loss of a vision—a vision of public spiritedness and concern for the collective welfare. The new nations of Africa were born in a moment of hope. And in an effort to transform their societies, political elites chose a mix of development strategies. Within the economic framework of these strategies enormously powerful private interests have entrenched themselves. The collective optimism of the nationalist era has given way to a sullen and embittered recognition that the sacrifices of the many have created disproportionate opportunities for the few. How do policy choices, ostensibly made for the public good, become the basis for private aggrandizement? By what process does a vision of the public order erode? This book investigates these critical questions.

A second theme is the role of the market as a political arena. In Africa, the market is the setting for the struggle between the peasant and the state. Through intervention in the market, the state seeks to manipulate the behavior of rural producers. It seeks to levy resources from the countryside: money, people, food, and raw materials. It also seeks to set terms for the supply of these resources, terms that restructure the patterns of advantage both within the countryside and between the countryside and urban industrial areas. For their part, the rural producers use the market as a means of defense against the state. By reallocating resources among economic alternatives, they seek to defend themselves against the depredations imposed upon them by many aspects of public policy. The result is a struggle in the marketplace between the peasants and the state, a conflict we shall discuss and analyze.

The marketplace is more than a locus of competition and conflict, however; it is also an instrument of political control. We shall show how government intervention in markets generates political resources, and how these resources are then distributed to build organized support for the political elites and the policies they propound. Market intervention becomes a basis for political control, and the way in which it does this goes far toward explaining some of the characteristic features of agricultural policies in Africa.

A last major theme concerns the broader fate of the peasantry in the development process. Barrington Moore once wrote: "Just what does modernization mean to peasants beyond the simple and brutal fact that sooner or later they are its victims?" (p. 467). As part of the modernization process, the peasantry is compelled to surrender its resources to the upper classes, to the state, and to the industrial sector. By looking at the attempts of the African states to transform their societies, this book will explore the ways in which governments and their allies seek to displace the peasantry, and to supplant it with a class of persons better suited to their conception of a modern industrial order.

In exploring these themes, we seek to contribute to several fields of scholarship. One, of course, is the study of agricultural development; another is African studies. In the ways just outlined, we seek also to advance the study of peasants. Standing alongside these fields is the literature in political economy, about which a few additional comments are in order.

Two variants of the literature in political economy are of particular relevance here. The first, which grows out of the study of agriculture, is exemplified in the work of such scholars as Michael Lipton and Keith Griffin. Although they have contributed path-breaking work, Lipton and Griffin tend to ignore African cases. And while outlining distortions in the development of agriculture introduced by political forces, they nonetheless concentrate more on documenting the consequences of political intervention than on analyzing why it takes particular forms. In this study, I attempt to further their work by extending it into Africa, and to deepen it by concentrating more fully on the political process behind the formation of agricultural policies.

A second variant of the literature in political economy focuses on the international setting of the Third World nations. Though fragmented in their ranks, scholars in this field share a common conviction: that patterns of change in Third World countries are largely determined by international political and economic forces, and that these forces originate in the industrialized nations.

It is clear enough, of course, that major forces affecting the prosperity of Africa have originated in the developed nations; the depression of the 1930s and the boom of the 1950s are the most vivid examples. But it is less well understood that African states strongly influence the specific ways in which these forces affect them. They do so out of regard for their own needs and the needs of powerful interests within their own societies. Furthermore, as I will show, the ways in which they manipulate these forces create enduring patterns of advantage within the emergent social order in Africa. This study therefore joins the work of others in arguing that to understand the patterns of development in Third World nations, scholars should pay more attention to the capacity for autonomous choice on the part of local actors, both public and private, and give greater weight to the importance of these choices in shaping the impact of external environments upon the structure of the local societies. (See Alavi, Warren, Swainson, Sklar, and Saul.)

This book falls into two parts. Part One (Chapters One through Four) is largely descriptive in nature. Part Two (Chapters Five through Seven) is largely interpretive. The arguments in both sections are advanced as hypotheses; though boldly stated, they are not proven. The book draws illustrative materials from southern Nigeria, Ghana, Kenya, Tanzania, Zambia, and the Sudan, and uses some additional sources from the Ivory Coast and the Sahelian countries, especially Senegal. None of the materials used come from the former "settler territories" of Southern Africa.

PART I

Government Interventions in Major Markets

CHAPTER 1

Policies Toward
Cash Crops for Export

The economies of tropical Africa are based on the production and export of primary products. In addition to such commodities as timber, minerals, and oil, African exports include agricultural products. Most important among them are the beverage crops—coffee, tea, and cocoa—and crops that yield vegetable oils: palm oil, palm kernel oil, cotton seed, and groundnuts. Also important are such fibers as sisal and cotton.

Like all nations in the developing world, the nations of Africa seek rapid development. Their people demand larger incomes and higher standards of living. Common sense, the evidence of history, and economic doctrine all communicate a single message: that these objectives can best be secured by shifting from economies based on the production of agricultural commodities to economies based on industry and manufacturing. The states of Africa, like states elsewhere in the developing world, therefore adopt policies that seek to divert resources from their "traditional" economic sectors (the production of cash crops for export) to their "modern" or "developing" sectors (their nascent industrial and manufacturing establishments).

A major factor that distinguishes many African states from others in the developing world is their possession of institutions for effecting this transfer. Most African states possess publicly sanctioned monopsonies for the purchase and export of agricultural goods. A monopsony is a single buyer; and where there are many sellers but only one buyer, the buyer can strongly influence the price at which economic transactions will take place. In Africa, public agencies are by law sanctioned to serve as sole buyers of major agricultural exports. These agencies, bequeathed to the governments of the independent states by their colonial predecessors, purchase cash crops for export at administratively determined domestic prices, and then sell them at the prevailing world market prices. By using their market power to keep the price paid to the farmer below the price set by the world market, they accumulate funds from the agricultural sector. Although the existence of international borders and the frequent absence of effective border controls have allowed some farmers to evade these state agencies, it has been estimated that at the time of independence, the agencies handled 90 percent of the exports of palm kernels, 80 percent of the exports of coffee, 65 percent of the exports of tea, and 60 percent of the exports of raw cotton (Temu, p. 12).

This chapter examines the market faced by the producers of export crops. It seeks to document the manner in which the governments have intervened in this market to transfer resources from the producers of cash crops to other sectors of society: the state itself; the new industrialists and manufacturers; and the bureaucracies that administer the market and manipulate the prices paid to farmers.

STATES AND THE REVENUE IMPERATIVE

The origins of the state marketing agencies (or marketing boards, the terms will be used interchangeably) lie in the colonial period.[1] Their individual histories are contrasting and complex; but they also

1. Most of the examples used in this book will be drawn from the English-speaking territories of West Africa. Materials contained in the studies sponsored by the Club du Sahel and the Center for Policy Studies confirm that the pattern prevails throughout much of Francophone Africa as well. More systematic evidence in support of the arguments is presented in Appendix B.

exhibit certain trends in common (for a recent review, see Jones 1980). The agencies were established in times of economic crisis, notably the Great Depression and the Second World War. And, almost invariably, they were officially mandated to use the bulk of the funds they accumulated for the benefit of the farming community.

Because agriculture represents the principal economic activity in most of Africa and often generates the greatest volume of foreign exchange, the agencies that controlled the market for agricultural exports soon became the wealthiest and economically most significant single units in their respective economies. Following World War II and the commodities boom of the 1950s, for example, many of them accumulated enormous reserves; as Bauer wrote in 1964, "their financial resources exceed those of the West African governments" (p. 263).

The marketing agencies are constrained by their enabling legislation to employ their reserves for specific purposes. When first established in the colonial period, they were mandated to devote the bulk of their funds to purposes beneficial to farmers. In Nigeria, for example, 70 percent of the trading surplus was consigned to the price stabilization fund. Portions of the remainder were to go to the development of the agricultural industry. In Nigeria, 7.5 percent of the reserves were to be spent in this manner; in East Africa, the percentage was much higher.

When confronted by the need for revenue, however, states have always found means of diverting funds from these agencies to the public coffers. During the 1940s, for example, the colonial governments used the marketing agencies to secure funds which they then "borrowed" at highly favorable rates of interest. In effect, this action compelled indigenous African farmers to subsidize the acquisition of war materials by their imperial overlords and the reconstruction of the homelands of their colonizers (see Hazelwood).

With the arrival of self-government in Africa, the revenue imperative strengthened. Like their colonial predecessors, the new states needed funds, particularly foreign exchange; unlike their colonial predecessors, they were deeply devoted to the development of their local economies. Thus they deliberately sought to transfer resources from agriculture to more "modern" activities in an effort to develop. Moreover, by contrast with the colonial regimes, the independent states of Africa were run not by appointed administrators but by elected politicians. With widespread politicization of the electorate in the nationalist era, politicians came under intense pressure from aggressive and demanding constituents. Those in control of the newly independent states therefore sought financial resources with which to reward the electorate. By comparison with the colonial period, the revenue imperative was thus stronger at the time of self-government in Africa. The result was that governments sought, and won, control over the revenues of the marketing agencies.

A vivid illustration of this process is offered by Obafemi Awolowo, an indefatigable figure in Nigerian politics. In the pre-independence political maneuvering in Nigeria, Awolowo and the Action Group, as his party was called, gained control of the Western Regional government. They did so by presenting a political program that promised lavish development expenditures, most notably on health and education. But when the Western Regional parliament opened, Awolowo and his Action Group government discovered that whereas the capital costs of their program would be £10 million, the total revenues available came to less than half that amount. As Awolowo wrote in 1960: "Where would the required money come from? That was the question. And it was a question which had to be tackled with speed and success, if we were to redeem our promises to the electorate" (Awolowo, p. 273).

Frustrated in various efforts to secure the needed funds, the Action Group imposed a series of new fees and taxes. The result was politically disastrous. The opposition "seized the opportunity to din it into the people's ears that they had been led up the garden path" (ibid., p. 275). In the federal elections that followed, the opposition campaigned on an anti-tax platform and captured a majority of the Western Region's seats in the Federal Parliament. The Action Group's plight became desperate. In the face of popular demands, Awolowo wrote, "we pressed on with our schemes" (ibid., p. 276). But how were they to secure the necessary funds? They did so, in Awolowo's words, through a "miracle," and the nature of the miracle is instructive.

The party that had defeated the Action Group in the federal elections itself held power in a regional government, the government of the Eastern Region. And it, too, was subject to popular pressures to furnish public services. The leaders of the rival parties therefore joined together in a coalition to resolve their common political dilemma; they formed an alliance and seized control of the marketing boards from the Federal Government, which had accumulated enormous resources from the trade in export commodities. As Awolowo wrote:

The real miracle occurred . . . when as a result of the alliance between the Action Group and the NCNC [its principal opposition] the Commodity Marketing Boards which were controlled by the Federal Government were regionalized, and allocation of revenue was made mainly in accordance with the principle of derivation. By means of the former, an accumulated reserve of over £34 million was transferred to the Western Region, and as a result of the latter our revenue rose from £6.39 in 1953–1954 to £13.20 million in 1954–1955. . . . Since the introduction of these financial measures, our revenue has been on a steady increase. [Ibid., p. 276]

The story told by Awolowo stands for a general trend in Africa. As public bodies, the marketing boards derive their powers from official statutes, and these statutes can be—and repeatedly have been—revised to make the boards more faithful servants of government. In particular, rather than being used to accumulate funds for the farmers, the agencies are increasingly used to impose taxes upon them.

This trend is illustrated by the role of these agencies in stabilizing, or failing to stabilize, prices. A major test of the intentions of the newly independent governments occurred almost immediately after independence, for between the crop years 1959–1960 and 1961–1962, the world price of cocoa fell approximately £50 a ton. If the resources generated by the marketing agencies were to be used to stabilize prices, then surely this was the time to use the funds for that purpose. Instead of stabilizing producer prices, however, the governments of both Ghana and Nigeria passed on the full burden of the drop in price to the producers. Under pressure from their governments, the marketing agencies, rather than protecting the producers, acted instead to stabilize the magnitude of the surpluses they accumulated from them—and as will soon be noted, they increasingly transferred these surpluses to their governments. The evidence tabulated in Appendix B demonstrates that this incident exemplifies a general trend. If the agencies were in fact following a policy of price stabilization, then it would be reasonable to expect that upon occasion they would have had to support domestic farm prices at levels in excess of the world price. But figures greater than 100 percent rarely appear in these tables.

Not only did the states ignore the legislated purpose of the funds; in efforts to secure revenues, they also altered the legislation. As Beckman notes, after independence in Ghana,

The government decided to remove certain legal restrictions on its access to the funds of the [Marketing] Board. Existing laws assumed that the funds were administered for the benefit of the cocoa-farming communities. The main purpose was price stabilization but development expenditure to meet their needs was also sanctioned. The Board was supposed to act as a trustee for the farmers. . . .

The government wanted to use the accumulated funds of the Board for its development program, without such . . . sectional restrictions. Legislation to that intent was presented in the National Assembly in July 1957. The Minister introducing the Bill declared that the cocoa funds 'should properly be regarded as being held in trust for all the people of Ghana.' [Beckman, p. 199]

A similar transformation took place in the Western Region of Nigeria. There, 70 percent of the trading surplus of the marketing board was to go for price stabilization, 7.5 percent for agricultural research, and the remaining 22.5 percent for general development purposes. But Helleiner (1966) notes that following self-government:

The Western Region's 1955–1960 development plan announced . . . abandonment of the "70–22.5–7.5" formula for distribution of the Western Board's trading surplus, offered a strong defense of the Marketing Board's right to contribute to development, and provided for £20 million in loans and grants to come from the Board for the use of the Regional Government during the plan. . . .

[The Board] was now obviously intended to run a trading surplus to finance the Regional Government's program. The Western Region Marketing Board had by now become . . . a fiscal arm of the Western Nigerian Government. [1966, pp. 170, 171]

This trend has also been noted in Senegal (IBRD, 1974), and more recently in the Ivory Coast, where the Caisse de Stabilisation (Stabilization Bank) diverts an increasing share of its funds from the stabilization of agricultural prices and the diversification of production to the capital investment fund of the national government (*West Africa*, April 28, 1980).

The loaning of money is thus one means by which the marketing agencies have transferred resources from the farmers to the state. The evidence suggests, however, that as time has passed governments have borrowed less frequently and taxed more often. Some "loans" are never repaid; others have been contracted at interest rates that range from 0 to 8 percent, in times when capital could rarely be secured for less than 18 percent in the private market (see Walker and Ehrlich, p. 340; Beckman, p. 204). Moreover, in Nigeria the regional loan boards have made fewer loans and more outright grants to their respective governments. By 1961 the value of grants exceeded that of loans, and by 1968 the transition was complete; as noted by Onitiri and Olatunbosun, "loans outstanding [to the government], which in 1961 were outstanding features of the Boards' investment portfolio, had completely disappeared. In their place, grants [to the government] have more than doubled" (p. 191).

Through the intermediary of the marketing boards, governments in Africa thus appropriate resources from export agriculture. The limited data available suggest that in the budgets of African nations, export agriculture commonly contributes from 20 percent, as in the cases of Ghana and Senegal (Morrison; Amin) to 40 percent, as in the case of Western Nigeria (Onitiri and Olatunbosun). In some cases, such as Uganda in the 1950s (Walker and Ehrlich), the figure is as high as 90 percent; and in others, such as Kenya in the 1960s (Sharply 1976), it is as low as 10 percent. But even the Ivory Coast, which has traditionally secured investment capital in the form of loans from abroad rather than in the form of levies from its farmers, now increasingly secures such capital from its agricultural price stabilization funds (*West Africa*, April 28, 1980).

Without knowing the allocation of government expenditures, we cannot say whether this level of taxation represents a redistribution of income. Unfortunately, data on the allocation of government expenditures is even harder to find than data on the taxation of agriculture. What little can be found, however, tends to indicate strong tendencies toward economic redistribution.

Reporting on a study of Ghana which he helped to conduct for the International Labor Organization, Ewusi writes:

[We] adopted the following means of estimating the size of capital formation by the government in the rural sector. The latest issues of the Annual Estimates of Government Expenditure are classified according to regions, [and] all forms of capital expenditure are shown with respect to the town, city, or village where the investment is located. Thus we summed up the capital investment in places which had a population of less than 5,000 and classified them as public investment in the rural areas. . . . The conclusion . . . is that the Government spends less than 5 percent of its capital expenditures in the rural sector. [Ewusi 1977, p. 90]

In an analogous study of development expenditures in Zambia, I found that well over 60 percent of the capital projects initiated in the first five years after independence were located in the urban areas (Bates 1976, p. 105). And in their analysis of the contribution of agriculture to the public revenues of Uganda, Walker and Ehrlich (1959) show that investments in hydroelectric power represented one of the major uses of public development funds. These investments were financed by "loan" funds from the marketing boards; but the primary beneficiaries of these expenditures were the budding group of industries in and about the major towns, and particularly the new industrial center of Jinja. A similar disparity is suggested in the figures for Tanzania. With less than 10 percent of its population in towns, its urban centers nonetheless secured 30 percent of the public expenditures under the state's first and second development plans (Clark, p. 98).

Besides revenues, the states of Africa need foreign exchange. As a leading sector of Africa's pre-industrial economies, export agriculture generates both revenue and foreign exchange. Using the price-setting power of the monopsonistic marketing agencies, the states have therefore made the producers of cash crops a significant part of their tax base, and have taken resources from them without compensation in the form of interest payments or of goods and services returned.

BUSINESS AND INDUSTRY: THE HEIRS OF THE NEW ORDER

In the front ranks of the intended beneficiaries of the redistribution of income from export agriculture stand the investors in industry and manufacturing. In part, this is by design: manufacturing is equated with modernity. In part it is a response to political influence: businessmen seek funds with which to establish enterprises, industrialists seek concessionary prices for raw materials, and both use instruments of the state to secure their needs by appropriating resources from the peasants.

State-Sponsored Capitalism

One of the best illustrations of the diversion of capital from agriculture to industry is provided by the materials from Western Nigeria. The government of Western Nigeria directly invested in promising industrial projects and also provided capital for investments by private individuals. The instruments for these two activities were two statutory corporations, the Western Nigeria Development Corporation and the Western Region Finance Corporation. The government provided the capital for both agencies. What is significant is the source of this capital and the terms on which it was made available. The source was agriculture, and the terms were concessionary.

During the period in which these corporations functioned they received the bulk of their finances from the Western Region Marketing Board. We lack detailed figures for the Finance Corporation, but we do know that between 1949 and 1958 the Development Corporation received £11.0 million from the Marketing Board (Oluwasanmi, p. 182), and that over 70 percent of its funds came from the Marketing Board (ibid., p. 129). The Board had little choice in the matter, because during this period it was under the supervision of the Ministry of Trade and Industry (Nigeria 1962, vol. 1, p. 37), and the Ministry diverted resources from the Board and into the promotion of industrial projects. When either the Development Corporation or the Finance Corporation sought funds, its directors often simply bypassed the Marketing Board and approached the Ministry directly; the Ministry would then issue a directive to the Board, instructing it to loan the requisite funds to the corporation requesting them (ibid., vol. 1, pp. 37ff). The result was the creation of a spate of new industrial firms—including printing companies, cement works, a glass factory, textile plants, a leather works, and a plastics company—financed in large part by loans secured from the marketing agency.

The resources secured from the Marketing Board were obtained on exceptionally generous terms. Indeed, investigations reveal that in many instances the two corporations simply failed to repay the Marketing Board and were often heavily in arrears in their interest payments (ibid., vol. 3, p. 44). Where repayment was made, the corporations were often able to secure radical extensions in the payoff period (from 5 to 15 years) and reductions in interest charges (ibid., vol. 1, p. 63). Moreover, the corporations often failed to safeguard the funds of the Board. When loaning money to local investors, "no arrangements were made . . . for taking securities" (ibid., vol. 1, p. 63). Even when the corporations did purchase securities, they often purchased nonparticipating shares, thereby failing to gain representation on the boards of directors of the enterprises and thus foregoing influence over the use of their funds (ibid., vol. 2, p. 1).

Because, in essence, the Marketing Board had to loan funds to the corporations, and because the corporations so thoroughly abused their privileged financial relationships, the Board thus became a means of redistributing income from agriculture to industry.

Local Industry

There is a second kind of firm that seeks privileged access to the resources of farmers: the firm that processes agricultural products. Firms of this type seek raw materials. And in their efforts to maximize profits, they seek the power to set the prices they pay to the farmers who supply them. For their part, the states of Africa seek to promote the development of these firms. Such enterprises offer a natural means of moving from an agricultural to an industrial economy. By processing agricultural products that have previously been exported for processing abroad, they also promise a means of retaining greater levels of "value added" within the domestic economy. The importance of these enterprises is affirmed both by conventional economists, who seek to increase forward and backward linkages, and by radical economists, who seek to lessen the dependence of poor countries upon international markets. States therefore promote the formation of these firms, and they do so in part by offering the prospect of low prices for raw materials. With the growth of local processing industries, then, investors and the state, whatever their differences may be (and sometimes they are major ones), form a political and economic alliance against the producers of cash crops. Let us consider how such alliances have affected three crops—coconut oil in Ghana and coffee and sisal in Tanzania.

Ghanaian Coconut Oil. The Esiama Oil Mill in Ghana, a large copra processing plant, was constructed in 1961 and designed to refine and export coconut oil. There were several rationales for construction of the mill, but perhaps the most persuasive was the relative technical superiority of processing copra with modern equipment (see Table 1). Working from plans provided by foreign engineers, local management imported expensive, highly advanced mechanical processing equipment. Two years after building the plant, and despite experiencing major operating difficulties, the management radically expanded the plant's capacity. Interviewing company officials in the mid-1970s, James Obben determined that the principal reason for the expansion was "the persistent belief that the area possesses a prodigious capacity to produce copra far beyond the projected maximum intake capacity of the factory. This obviously derives from the strong impression obtained from observing miles of continuous stretches of [forests] in the area, which has been assumed to be a reliable index of real supplies" (p. 25).

Major problems soon developed, however. The layout and design of the plant proved defective and the mechanical equipment

Table 1
Efficiency of Traditional and Industrial
Methods of Coconut Oil Extraction

Methods of processing	Average oil content (percent)	Rate of oil extraction (percent)	Oil remaining in cake (percent)
Traditional methods	67	38–45	22–29
Industrial methods	67	56	12

Source. James Obben. *A Study on the Costs of Processing Coconut at the Esiama Oil Mill and the Economic Viability of the Venture*. Dissertation submitted to the Department of Agricultural Economics, University of Ghana, June 1976, p. 19.

proved unreliable, particularly under local operating conditions. In the economic environment prevalent in Ghana, repairs were difficult; a breakdown in the machines could take two to three months to repair, a fact which cut deeply into production (Obben, p. 20).

The technical superiority of industrial methods of coconut oil extraction thus failed to provide an accurate guide to their relative economic merits. For given the capital-intensive technology of the plant and the frequency and extent to which the equipment stood idle, the plant could produce oil only at very high costs. In 1975, for example, its unit costs of production were 987.98 *cedi* per ton; the value of its production on the international market (c.i.f.) was ₵624.89 (Obben, p. iv).

To lower its costs, the management therefore attempted to secure its raw materials at reduced prices. The price it offered lay below that offered by the "traditional processors" of copra oils, however, and the firm was therefore frustrated in its attempt to secure adequate supplies. In the end, it had to secure what amounted to a charter to serve as a monopsonistic buyer of the output of local producers; in effect, it was empowered to form its own marketing board. And with the backing of the police powers of the state, the firm excluded competitors seeking to bid for the copra crop.

Tanzanian Coffee. Coffee growers in Tanzania are paid for their products by a crop authority which acts as a government-sponsored monopsony. The price that growers receive for their products is determined by the final selling price adjusted for the costs incurred by the crop authority. A cursory examination of the crop authority's costs reveals that the biggest single share is one designated as "local roasting subsidy." This subsidy is another public policy measure designed to promote the development of domestic firms that process agricultural products.

The government of Tanzania has sought to take advantage of the local production of coffee by establishing a firm to manufacture soluble coffee for sale in the markets of East Africa. To promote the development of this firm, the government has mandated that sales of coffee shall be made to it at prices below the world market price. Whereas robusta coffee commanded a price on the world market of Tshs 14.84 per kilo in 1975–1976 (Tanzania shillings), the local manufacturer could purchase it from the crop authority at Tshs 6.32 per kilo. Had all sales been made at the world market price, the coffee authority would have increased its earnings by Tshs 16 million, according to one government source; and of course the producers would then have realized higher prices for their crop (Tanzania 1977e).

Sisal in Tanzania. Whereas coffee is grown by small-scale producers, sisal is grown on large estates. Nonetheless, the evidence from Tanzania suggests that governments are willing to sacrifice the interests of even large-scale producers in efforts to construct manufacturing capabilities.

Since the Arusha Declaration of 1967, the government of Tanzania has attempted to move from the export of raw materials to the export of processed goods. To secure this objective, the government has sought to create a domestic capability for the manufacturing of rope and twine. In this it has succeeded. There now exist six major spinning mills in Tanzania with a capacity to process 115,000 tons of sisal annually. Projections suggest that by 1980 the country will possess the capacity to process over 90 percent of its domestic sisal production. Furthermore, whereas in 1967 Tanzania exported less than 11 percent of the value of its sisal exports in processed form, by 1976 over 30 percent of that value was in the form of finished products (Tanzania, 1977d).

The development of sisal manufacturing in Tanzania has been financed in part by the sisal producers, and both export taxes and pricing policy have been used to reallocate resources from the producers to the processors of that crop. Sisal is subject to export taxes; in 1974 and 1975, revenues from this tax amounted to over 100 million Tanzania shillings. By statute, 50 percent of the tax revenues are paid into a special "sisal products fund" and thus earmarked for the development of the sisal industry. Payments from this fund are governed by the Tanzanian treasury, and government reports indicate that "most of the proceeds of the fund to date have been used to finance investments in sisal spinning" (Tanzania 1977d, p. 29).

The transfer of revenues from producers to manufacturers is also promoted through pricing policy. The monopsony buyer of the sisal crop—the Tanzania Sisal Authority—sells essentially to two consumers: the "world market" and domestic manufacturers. The Sisal Authority has chosen to make its sales to the domestic manufacturers at a price well below the world market price for raw sisal fiber. And because the Authority pays the farmers the residual difference between the sales price and its costs of marketing, the result of selling at a reduced price to domestic manufacturers is to lower the price paid to producers of the crop.

In 1977, the Ministry of Agriculture reported: "In 1976, the Tanzania Sisal Authority sold 36,072 tons of fiber to local mills at an average price of Tsh 1984 per ton. In comparing this with an average export price of Tsh 3,007 per ton in 1976, account must be taken of differences in grades and the timing of sales." (Tanzania 1977d, p. 27). Despite its caution, this report insisted that "sales to local spinning companies have been heavily subsidized." It also made clear that it was the producers who bore the cost of this subsidy in the form of downward adjustments in producer prices—adjustments that reflected the lower average realization for sales by the Sisal Authority.

Similar cases abound: the refinement of palm oil by government mills in Eastern Nigeria (Kilby; Usoro); the operation of plants to produce cocoa butter and cocoa powder in Ghana and Nigeria (Killick; Schatz 1977); the conduct of sugar estates in Ghana (Killick), Del Monte's pineapple cannery in Kenya (Swainson 1977a), and

INDECO's cannery in Zambia (Baylies 1978); the operations of the cotton mills in the Ivory Coast (Campbell 1974); and the behavior of the vegetable canning corporations in Ghana and the vegetable oil firms in Sudan (Grayson; *African Business*, February 1980). All of these entailed depressing the prices paid to the producers of cash crops in an effort to help finance the formation of domestic manufacturing firms.[2]

These examples illustrate that governments in Africa are willing to sacrifice the interests of farmers in order to promote the formation of industrial establishments. They do *not* demonstrate, however, that governments are willing to compromise *any* interest to safeguard industrial profits. To the contrary, governments are often willing to lower the profits of firms in order to secure other objectives—such as a plant location that is politically desirable though economically disadvantageous, or the maintenance of a labor force that is too large to generate maximum profits. What these examples do illustrate, and what is important here, is that governments are willing to undercut the interests of rural producers to promote the development of industry.

The development of local manufacturing establishments is an important achievement—perhaps even a watershed in African history—but it is not one that the farmers necessarily welcome. Processing formerly took place in the advanced industrial nations, and Africa's economic role was confined to the production of raw materials. As we have seen, this is rapidly changing. Increasingly, Africa possesses the capacity to transform raw materials into finished or intermediate products. But the interests of African farmers are

2. As Schatz states in his discussion of the growth of manufacturing in Nigeria: "Processing operations were sometimes established with the inducement of substantial subsidization through the privilege of purchasing Marketing Board export crops at prices below the world market level. In a number of cases, this subsidization exceeded the world market value added by domestic processing, so that 'effective subsidization' . . . exceed 100 percent" (Schatz 1977, p. 125). Killick cites a particularly gruesome case, originally reported by Norman Uphoff. The Ghanaian government had built a tomato-canning plant. In order to test the plant, the management brought in the police and border guards to keep away private buyers while the management bought up the thirty tons of tomatoes required for a test run of the factory (Killick 1978, p. 233).

often sacrificed in securing this transition. In the late 1970s, for example, the Federal Government of Nigeria banned the export of groundnuts to the international market (*African Business*, May 1979). It did this in an effort to secure adequate supplies of raw materials for the local crushing industry at prices the industrialists could afford. The interests of local industry thus led to a restructuring of the market by the state, and a historical marketing pattern was broken. But in this transformation, the farmers bore a major portion of the costs.

The Bureaucracy

The state and industry are not the only beneficiaries of this transfer of resources from agriculture to other sectors, however. Another is the bureaucracy, which organizes the market and is charged with manipulating it for public purposes.

Some evidence of the magnitude of the resources that accrue to the bureaucracy is the size of the costs of marketing. In a study of capital flows out of agriculture in Kenya, Jennifer Sharpley (1976) found that marketing costs accounted for between 10 to 35 percent of the differential between the world market price and the price paid to domestic producers (p. 110). As she states: "In 1969, of the various financial adjustments that could be estimated, marketing costs . . . were the largest item. . . . Financial transfers through taxation, subsidies, loans, and direct investment were found to be considerably smaller in size" (p. 109).

More recent research also reports a sharp inflation in the Kenyan costs of marketing. Over the period 1971–1976, the unit costs of marketing coffee increased 32 percent, of wattle bark 44 percent, and of cotton 406 percent (Gray, p. 64). Indeed, expenses have increased so greatly that a special review committee has called for a reform of the marketing boards (*Weekly Review*, June 6, 1979). Nor is the problem confined to Kenya. Whereas marketing costs consumed 7.4 percent of the value of cocoa sales in Ghana in the 1950s, by the late 1960s they had risen to over 17 percent. A similar pattern prevails in the cocoa industry of Nigeria (Beckman; Wells, p. 204).

In part, the rising costs of marketing result from plain inefficiency: poor storage, inefficiently scheduled transport and disposal of the crop, and careless contracting in both procurement and sales. All these problems bedevil the marketing boards, and all appear to be exacerbated by their monopsonistic status: because they are able to set prices, they can afford to be inefficient, for they can pass the costs of their inefficiency on to the farmers. Their inefficiency takes another form: growth in the number of their staff members and the perquisites they receive. The best evidence of such tendencies comes from Ghana, where one commission of inquiry noted:

The evidence before us suggests that the [Cocoa Marketing Board] used the profits obtained from its monopoly cocoa operations to . . . provide funds for the dance band, footballers, actors and actresses, and a whole host of satellite units and individuals. . . . the State Cocoa Marketing Board itself is not free from . . . this type of practice. The CMB's area of operation . . . embraces activities and involves a staff which would have appeared absurd only ten years ago. [Ghana 1967a, p. 28]

This commission also noted the ability of marketing personnel to abuse their monopsonistic positions so as to radically enhance their incomes:

Farmers often referred to the opulence of the Secretary Receiver [the official who operates the local buying station]. It was alleged that these officers who earned £G180 per annum owned cars, trucks, buildings, etc., and often supported as many as three wives. We saw some Secretary Receivers owning Mercedes Benz cars, Peugeot cars, and transport trucks. [Ghana 1967a, p. 20]

Similar abuses pervade the upper levels of the bureaucracy as well. Thus, recent inquiries into the Cocoa Marketing Board suggest the extent to which the directors of the board divert the trading surpluses accumulated from farmers to their own pockets. *West Africa* reported:

Cmdr. Addo, former chief executive of the Cocoa Marketing Board, told the committee investigating its affairs that the CMB spent nearly ₵1 m. on drinks alone between August 1977 and July 1, 1978. Giving evidence, Cmdr. Addo said during his tenure of office he instituted certain measures to boost the morale of the directors. As part of these measures, he said, all the eight or ten directors were given a bottle each of whisky, brandy, and gin at the end of every month in addition to receiving a . . . table allowance. [*West Africa*, Nov. 27, 1978, p. 2386]

In addition, Commander Addo stands accused of fraudulently appropriating hundreds of thousands of *cedis* of the Board's trading profits (*African Business*, January 1980).

The tendency to use marketing channels to appropriate revenues generated by the production of cash crops is not confined to civil servants. Similar tendencies have arisen when the state has empowered cooperative societies to serve as marketing channels. This has been most thoroughly documented in Tanzania, where investigations in 1966 (Tanzania 1966) and 1970 (Kriesel et al.) disclosed rapidly inflating marketing costs on the part of cooperatives, and specified the number and the emoluments of their staffs as major causes of this trend.

In Kenya, where cooperatives have been retained in many sectors of the agricultural industry, an alarming increase in marketing costs has also taken place. In a recent report the International Coffee Organization wrote: "It will be noted that deductions by cooperative federations and cooperative societies have increased from 17.3 U.S. cents per pound in 1974–75 (U.S. $23 per bag) to 36.3 U.S. cents per pound (U.S. $48 per bag) in 1975–76 and 50.9 U.S. cents per pound (U.S. $67 per bag) in 1976–77" (ICO 1978a, p. 24). Partly as a result of these deductions, the small-scale coffee producers, who market through the cooperative societies, obtain roughly 30 percent less of the portion of the world market price for coffee received by the large-scale plantations, who market directly through the coffee board.

CONCLUSIONS

We have examined here the position of the producers of cash crops for export. We have seen that they have been subject to a pricing policy that reduces the prices they receive to a level well below world market prices. And we have noted that although some of the resources expropriated from agriculture are returned in the form of interest payments and public services, perhaps most of these resources are diverted to other sectors—to the state, to urban-based industrial enterprises, and to the bureaucrats who administer the publicly structured markets for farm products.

The tabulation in Appendix B documents the level of the financial burden placed on the producers of export crops by the dual price policy of the public marketing agencies. In most cases, the data represent the prices offered to domestic producers expressed as a percent of the f.o.b. price at the nearest major port. In some cases, they represent the percent of the income generated by the sale of the crops abroad that is actually secured by the producers. In either case, Appendix B shows that the producers have almost invariably received a price lower than the world market price. In most instances, they obtained less than two-thirds of the potential sales realization, and in many cases they received less than one-half.

CHAPTER 2

The Food Sector:

The Political Dynamics of Pricing Policies

Political pressures for low-cost food come from two main sources. One, of course, is the urban worker. The other is the employer who, when his workers are faced with high-cost food, is forced to pay higher wages. For political reasons, African governments must appease the urban worker; but as major employers and as the sponsors of industry, governments share the interests of those who pay the wage bill. To appease consumers while pursuing their own interests, governments therefore join with workers and industry in seeking low-cost food.

The issue that most frequently drives African city dwellers to militant action is the erosion of their purchasing power. The force of consumer interests was clearly revealed in the nationalist period. Following the Second World War, a combination of worldwide inflation and the resistance of colonial firms and governments to claims for offsetting wage increases led to widespread protests throughout the urban areas of Africa. In Ghana, for example, the anti-inflation campaign organized among the urban consumers gave strong impetus to the nationalist movement (for example, see Austin, p. 71ff), and a similar story can be told for other territories (see the studies in Sandbrook and Cohen; Berg; and Gutkind et al.). By capitalizing on the political disaffection engendered by inflation, nationalist politicians seriously weakened the power of the colonial administrations and significantly hastened their own rise to power.

Since independence the militance of the urban consumers has remained largely unabated. The contemporary histories of many of the independent African nations might credibly be recorded by focusing on major periods of strike action and worker protest; on major wage concessions by state corporations, public services, or private industry; and on the work of major tribunals or commissions of inquiry into labor unrest. The Turner Commissions in Zambia and Tanzania; the Brown Commission in Zambia; the Gorsuch Commission, the Morgan Commission, the Adebo Commission, and the Udoji Commission in Nigeria; the unrest which led to these commissions, the reports they issued, and the government white papers issued in rejoinder—all these bear witness to the continued importance of urban demands for higher standards of living.

Not only have consumer interests remained militant; governments have remained vulnerable to consumer disaffection. The colonial regimes were not the last governments to lose power in part because of increases in the cost of living. The fate of the Busia government in Ghana is illustrative. For a variety of reasons, in December 1971, the Busia government decided to devalue the *cedi*. The result of the devaluation was an immediate rise in prices, not only of imported items but also of locally manufactured items that faced reduced competition from abroad. As reported by Libby: "The *Ghanaian Times* announced that, according to a survey conducted by the official Ghana News Agency on December 30, a packet of St. Louis sugar which formerly sold at . . . 27 New Pesewas now sold for 40 NP; a tin of Peak milk sold at 15 NP instead of 11 NP. On January 6, 1972, a can of beef sold for 70 NP instead of 55 NP and a packet of locally manufactured cigarettes sold for 65 NP instead of 45 NP" (pp. 85–86). The rise in prices sparked widespread discontent; and in response to the ensuing strikes, demonstrations, and public disturbances, the military seized power. As Libby contends: "The public reaction to devaluation was sharp and hostile. It created a climate in which a military coup d'etat could be carried out" (p. 86). Subsequent price rises in Ghana helped provoke subsequent coups. And the withdrawal of the military from power in the late 1970s was accelerated by the wave of strikes in 1978, and by the military's inability to assuage the economic grievances of the urban workers, and its proven inability to force them, in the face of eroding standards of living, to provide labor services. Sadat, Nimeiri, Kaunda, Moi, Gowan, and Tolbert are among the other African leaders whose governments have felt the political pressures generated by the erosion of the purchasing power of urban dwellers; in the face of these pressures, several have fallen.

African governments can respond to demands for higher real incomes in several ways. But the choices they make are shaped by the fact that they, too, pay a bill for wages. Beyond paying civil servants and bureaucrats, in most cases they must also pay those who operate the ports and harbors, the railways, and the national transport systems. By establishing new industries or nationalizing existing ones, governments have become the owners of firms. They have also formed partnerships with private investors, thus becoming the owners of large-scale enterprises. And, hungry for capital to promote further investments, many governments strive to maintain an attractive environment for foreign investors. For all of these reasons, governments in Africa tend to resist demands for higher wages; they look for other alternatives.

One of the options available to political leaders, of course, is to attempt to reduce the effectiveness of organizations that seek to advance the economic interests of urban workers. One tactic they have used is co-optation: appointing labor officials to government boards, directorates in public enterprises, and central committees of governing parties; providing lavish office buildings and other perquisites; and providing for compulsory checkoffs and other measures that serve the organizational interests of trade-union movements. An alternate tactic is to suppress trade unions. By banning strikes, jailing labor leaders, and dissolving unions or compelling them to merge into government-sanctioned labor movements, they have sought to cripple the power of organized labor to champion the interests of urban consumers.[1]

1. There is a large, though uneven, literature on African labor movements. For reviews of it, see Kraus and Friedland. The best recent books are by Cohen, Gut-

Policies that seek to curtail urban demands by crippling their organized expression are only partially successful, however. Direct attacks on labor movements are open to reprisals; in moments of economic stress, labor movements can join with their urban constituents, paralyze cities, and create the conditions under which ambitious rivals can displace those in power. And attempts at co-optation still leave open the chance for wildcat actions; during moments of economic crisis in the cities, workers can and have acted on their own, and elite-level champions have been willing to come forth to lead them.

Thus governments face a dilemma: urban unrest, which they cannot successfully eradicate through co-optation or repression, poses a serious challenge to their interests as employers and sponsors of industry. Their response has been to try to appease urban interests not by offering higher money wages but by advocating policies aimed at reducing the cost of living, and in particular the cost of food. Agricultural policy thus becomes a byproduct of political relations between governments and their urban constituents.

The relationship between urban unrest and agricultural policy is an immediate one. When the Busia government of Ghana was overthrown in 1972, one of the first acts of the new military government was to publicly champion Operation Feed Yourself, a package of programs designed to secure greater food production and lower urban food prices. The Easter Rebellion in Liberia in 1979, which ultimately led to the overthrow of the Tolbert regime, eventually produced a basic change in agricultural policy. A central issue in the

---

kind et al., Jeffries, and Sandbrook and Cohen. The interchanges between Berg and Meeks are important. See also the debates between the devotees of the labor aristocracy thesis of Fanon, Arrighi, and others and the adherents to the more classically Marxist position. The debate is aptly summarized by Sandbrook (1977), among others. One of the leading scholars of the political role of urbanites, Joan Nelson, has found that both social scientists and policy-makers tend to overstate the radical tendencies of urban dwellers in the developing areas. Nelson does stress, however, that although the urban poor may not participate militantly on behalf of radical platforms, they do enter the political arena in pursuit of immediate, concrete economic gains (Nelson, pp. 138ff). Nelson also stresses that food prices are a central interest, and that urban demands for low-cost food result in policies deleterious to agrarian populations (ibid., pp. 343ff).

rebellion was the urban cost of living, and in particular, the government's announced intention of raising the price of rice. The immediate results of the rioting were the arrest and detention of opposition groups which had sought to capitalize on urban discontent, and a Presidential decree revoking the decision to raise rice prices and proposing instead a series of producer subsidies designed to elicit greater rice production.[2] President Tolbert was later overthrown; the changes in agricultural policy remain in effect.

This pattern also pervades policy-making in Nigeria. Following the defeat of Biafra in 1970, worker unrest, long suppressed during the civil war, led to widespread work stoppages. In response, the government convened the Adebo Commission, which in August 1971 gave an award of 12 to 30 percent wage increases to all members of the public service, and publicly called for "adjustments . . . to be made to wages and salaries in the private sector" (Nigeria 1975b, p. 16). Workers in the private sector enthusiastically championed the extension of "Adebo" awards beyond the public services. The results were across-the-board wage and salary adjustments.

Coupled with higher levels of government spending in the early 1970s, this increase in wages and salaries led to further price rises and to renewed demands for wage increases. The government convened another commission. Basing its recommendations on the need to adjust wages and salaries to rising price levels, the commission recommended "substantial salary increases for most grades of workers in the public service, ranging in most cases from 8 percent to more than 100 percent" (ibid., p. 17). Reporting in September 1974, the commission backdated its awards to April 1974. "The resulting arrears, paid between January and February 1975, pumped a vast sum of money into circulation" (ibid., p. 75).

Both commissions, in dealing with the problem of urban discontent, introduced a new emphasis in the government's policies toward urban radicalism: an increasing determination to deal with the problem not only by increasing urban incomes but also by curing its apparent cause—the rising cost of consumer items. As the Adebo

2. See the accounts in *Africa*, No. 94 (June 1979), and *African News*, April 27, 1979, and June 8, 1979.

Commission stated: "It was clear to us that, unless certain remedial steps were taken and actively pursued, a pay award would have little or no meaning and could indeed make matters worse. Hence our extraordinary preoccupation with the causes of the cost of living situation" (Nigeria 1971, p. 9). And as part of its effort to confront the causes of the rising cost of living, the commission went on to recommend a number of basic measures, among them many proposals designed "to improve the food supply situation" (ibid., p. 10).

In June of 1975, following the overthrow of the Gowon regime—whose unpopularity was caused in part by its apparent inability to deal with rapidly rising consumer prices—the new government of Nigeria appointed a task force to investigate the problem of inflation. This body, too, pinpointed the need to increase food supplies and reduce food prices as a key element in any attempt to assuage the demands of the urban working population. (See Nigeria, 1975a and 1975b.) Largely in response to the recommendations of this commission, the government of Nigeria adopted a series of highly publicized policy measures to increase agricultural production. This mixture of policies, which will be discussed further below, was commonly referred to as Operation Feed the Nation.

Agricultural policy is thus derivative. It is devised to cope with political problems whose immediate origins lie outside of the agricultural sector. Pricing policy finds its origins in the struggle between urban interests and their governments; and in the political reconciliation of that struggle, it is the rural producers who bear the costs: they are the ones who bear the burden of policies designed to lower the price of food. African governments attempt directly to alter food prices in two major ways: through the manipulation of trade policies, and through the operation of government-controlled marketing institutions.

COMMERCIAL POLICY

An exchange rate is the rate at which one currency can be exchanged for another. When a government appreciates the value of its currency—for convenience, we may call that currency the "dollar"—then the holders of dollars need pay less in order to secure a given amount of another currency. In effect, the government increases the worth of its "dollars." Following official measures to appreciate the value of a national currency, citizens then find that the price in "dollars" of foreign goods is lower than before. Foreign goods appear to be cheaper. Consumers can import them at a lower dollar price from abroad; and producers, unless the government gives them tariff protection, find the price of goods sold by their foreign competitors to be lower than before the government's action.

For various reasons to be discussed in later chapters, governments in Africa, as in other developing areas, maintain overvalued exchange rates. In order to facilitate certain kinds of imports, they appreciate the value of their currencies above a level that would be warranted under free-market conditions. As a consequence, they reduce the domestic price of food. They do so by maintaining an overvalued exchange rate and by failing to adopt a structure of protective tariffs that would compensate for the resultant lowering of the perceived price of foreign food supplies. They also do so by allowing food to be imported when the domestic price exceeds the world price, and by banning its export when the opposite holds true.

Examples of these measures may be found everywhere in Africa, but some of the most apposite came as part of the anti-inflation policy package devised in Nigeria. One example concerns wheat. A World Bank mission to Nigeria in 1978 reported that imports of wheat had risen dramatically in the late 1970s. One reason, it noted, was that the price of bread had been fixed since January 1974. Urban incomes were rising, and as people became better off, they tended to switch to the consumption of bread prepared from wheat flour. The result was a rising demand for wheat. Moreover, the report continued, "at current exchange rates, wheat can be imported much more cheaply than it can be produced locally. Wheat can [also] be imported duty free" (IBRD 1978b, p. 12). Rice offers another example. Following the recommendation of the anti-inflation task force set up following the displacement of the Gowon regime, the Nigerian government reduced the duty on rice from 20 percent to 10 percent; with rising domestic prices, demand shifted to foreign sources, and given the overvaluation of Nigerian currency, imports rose over 700 percent. As the World Bank report concluded: "The overvalued exchange rate is consumer biased. The massive importation of rice and wheat keeps the price of these and substitute commodities lower than would occur under restricted imports or a lower exchange rate" (IBRD 1978b, p. 38).

Besides sometimes encouraging the importation of foreign crops when domestic prices lie above world market prices, governments sometimes ban the export of food crops to prevent local prices from rising to a higher international price. In December 1974, for example, the government of the Sudan imposed an export duty of 20 percent on meat and meat products, thereby making it unprofitable for domestic producers to sell on the growing Mideast market and lowering the price to domestic consumers. In Kenya, both the Kenya Meat Commission and the Kenya Cooperative Creameries are compelled by government regulations to offer their products on the internal market at prices well below the world price for meat and dairy products; the same is true for the Kenya Tea Authority. These agencies lobby strenuously to be freed of such controls; they are nonetheless compelled to supply the domestic market at prices below the world price and to suffer the resultant loss of profits (Gray; *Weekly Review*, December 1, 1978; interview, August 1978).

The manipulation of protective measures can strongly affect the economic fortunes of domestic producers. Again, one of the most striking illustrations comes from Nigeria. Seeking to improve urban diets by promoting the production of poultry, the government of Nigeria encouraged the development of new, high-yielding varieties of yellow maize to be used as feed. The government enlisted the assistance of the International Institute of Tropical Agriculture (IITA), a Nigerian-based, internationally supported agricultural research institute. The IITA, in an expensive, ingenious, and innovative program conducted in collaboration with the Nigerian government, developed a new variety of maize that was widely accepted by local farmers and promised to dramatically increase local feed supplies. By 1977, the IITA effort moved out of the pilot stage and began to be incorporated into the agricultural programs of several of the states of Nigeria; 2,500 villages were targeted to receive stations providing the new varieties and supporting technical assistance. Then the government reversed itself. In April of 1977, in an effort to lower urban food prices by cutting the production costs of poultry, it removed all barriers to the importation of yellow maize. With protection removed, the overvaluation of Nigeria's currency meant that U.S. number one corn could be imported at 150 *naira* a ton, over 100 *naira* a ton below the local costs of production. Local farmers who had adopted the IITA package found that they could no longer sell their maize at a profit, given prevailing prices in the Nigerian market. As a result, the experiment was devastated (interviews, July 1978).

OFFICIAL MARKETING CHANNELS

An alternative method of reducing prices is for the government to intervene directly in the market for food. In accord with policies analogous to those imposed on the producers of export crops, the governments of some countries have created legalized monopsonies in the form of marketing boards for foodstuffs. These agencies buy produce at officially mandated prices and sell food products through price-controlled channels in town. Examples of such government agencies would be the National Agricultural Marketing Board (NAM Board) in Zambia, the Maize and Produce Board in Kenya, and the National Milling Corporation (NMC) in Tanzania. Similar bodies are found throughout the countries of the Sahel (Club du Sahel; Center for Policy Alternatives).

Both the NAM Board and the NMC provide substantial subsidies to urban consumers. In 1973, for example, the NAM Board received credits of K13.5 million from the government treasury to support the official government maize price; in 1974, the programs cost the government K12 million (Dodge, p. 116). Although I lack similar figures for Tanzania, official sources freely acknowledge that the National Milling Corporation (NMC) operates at a loss, and that its efforts to maintain low retail prices are a major reason for this.

The Bureau of Marketing and Research of the Tanzanian Ministry of Agriculture commented in 1977: "Retail prices for the main cereals have not been increased for three years. . . . Although consumer prices are outside the scope of the Annual Price Review, the NMC's present precarious financial position indicates a clear need for a close examination of present cost and price structures" (Tanzania 1977a, p. 6). The tendency to incur debts in order to support low urban prices has also been noted in the countries of the Sahel (Club du Sahel).

In seeking to maintain low consumer prices, the marketing agencies attempt to increase urban food supplies. They do so by importing food from abroad and distributing it in the urban market. Government-sponsored food imports have become a regular feature of the agricultural cycle in Africa: as the planting season begins and domestic food stocks dwindle, African governments enter the world market in search of food. And by importing food, the marketing agencies in effect compete with the local farmers in supplying the urban market, thereby lowering the price of the farmers' products (see the discussion in Club du Sahel, pp. 40ff).

The marketing agencies also seek to lower the price of food by lowering the prices they offer the farmer. In Tanzania, for example, between 1971 and 1976, the government offered prices that ranged from one-fifth to one-half the world price (Tanzania 1977c, Table 2.4). And in Zambia, Dodge found that if maize producers had been able to sell on the world market instead of to the NAM Board, they could have nearly doubled the prices they received in 1970–1971 and secured prices 50 percent higher than they received in 1973–1974 (Dodge, p. 118).

To impose these prices on farmers, governments establish a bureaucratic machinery to control marketing in the countryside. The regulations make the government the sole buyer of the crop. The prices are then set by law, and farmers who market their products outside official channels are subject to legal action. To control maize marketing in Kenya, for example, and thereby control the price of maize, the government requires anyone seeking to move more than ten bags of maize within a district, or two bags across district lines, to secure a movement permit. In this way it seeks to make the Maize and Produce Marketing Board the sole buyer of the vast bulk of the maize crop.

It should be stressed that government attempts to control the market for food crops have failed. By contrast with the market for export crops, the market for food crops is extremely difficult to control. Many export crops can be grown only in highly specialized areas, but food crops can be grown virtually by all farm families. And whereas export crops must be moved through a few special locations—ports, for example—food crops can be moved in many ways. Government policing of the marketing and distribution of food crops is therefore more costly. Moreover, export crops often require specialized buyers: persons with access to foreign markets or to very expensive means of processing. Food crops, by contrast, can be bought by almost anyone, and can generally be processed by the consumers themselves. As a consequence, food crops can more readily be diverted from official marketing channels.

There is strong evidence that these factors have rendered government efforts to lower the price of food in domestic markets less successful than efforts to lower the price of cash crops in export markets. In countries that have marketing boards for domestic foodstuffs, no more than 10 to 30 percent of the crops designated for government control actually pass through official channels (Kenya 1972; Temu, p. 172; also Jones 1972). Instead, the bulk of marketed production is distributed through "unofficial" channels. Furthermore, administrative controls have failed to restrain price increases. The urban consumer in Africa is suffering, and in large part because of increases in the price of food. (Tables 2 and 3 display the rise in food prices in Nigeria and Ghana.)

Clearly, the governments of Africa have failed to provide low-priced food by organizing the market for farm products. One consequence of this is that they have increasingly tried other methods, which we shall explore in the next chapter. Despite these other efforts, their marketing agencies remain in place. This is a testament to the power of urban interests. Given the political realities of contemporary Africa, it is extremely difficult for governments to

Table 2
Composite Consumer Price Indexes, Urban Areas in Nigeria, 1970–1976
(1960 = 100)

		Consumer price index	Food component of index
1970	March	145.2	154.0
	June	153.4	169.8
	September	155.0	172.4
	December	155.8	172.9
1971	March	166.5	194.8
	June	182.5	229.8
	September	177.4	216.6
	December	179.1	218.8
1972	March	184.7	227.8
	June	185.6	231.0
	September	173.6	204.0
	December	172.9	199.9
1973	March	182.6	211.7
	June	194.5	233.2
	September	189.6	221.4
	December	203.9	244.6
1974	March	206.2	246.8
	June	218.6	268.4
	September	219.6	262.1
	December	224.0	269.8
1975	March	259.2	322.3
	June	293.3	381.7
	September	303.7	398.3
	December	317.7	420.7
1976	March	336.7	451.9
	June	348.3	469.9
	September	355.3	466.3
	December	360.3	467.9

Source. International Bank for Reconstruction and Development. *Nigeria: An Informal Survey.* Lagos: Typescript, 1978, Table 3.

Table 3
Indexes of Retail Prices for Accra, Ghana
(1960 = 100)

	Local food prices	All retail prices
1960	100	100
1961	106	106
1962	117	116
1963	122	121
1964	137	134
1965	188	169
1966	216	192
1967	184	176
1968	201	189
1969	218	203
1970	228	208
1971	256	227

Source. Tony Killick. *Development Economics in Action*. New York: St. Martin's Press, 1978, p. 95.

terminate a program or to withdraw from the countryside a bureaucracy whose function is to secure lower food prices.

One major consequence of the persistence of these institutions is continuing conflict between peasant and bureaucrat in the rural markets of Africa. The peasants exploit the economic alternatives which the market offers in an effort to avoid the adverse impact of official policies. The bureaucrats seek to appropriate the peasants' products at the lower prices the state is willing to offer. Another effect is more subtle. Whereas at the level of official policy, the interests of the peasants and the bureaucrat are in conflict, at the level of unofficial practice they are often consonant, given the structure of the incentives to which the official policy gives rise. To put it bluntly, the policies offer joint gains through corruption. The bureaucrat can offer protection against the very policies he is mandated to impose: for a portion of the gains, he can help the peasant evade market controls. And the peasant, rather than attacking government policy directly, can often do better by seeking to become an individual exception to it; he can do this by offering bribes. Within the pattern of conflict to which government market intervention gives rise, this style of accommodation between the bureaucrat and the farmer becomes an important means by which African governments evoke individual compliance with policies that are collectively harmful.

CONCLUSIONS

Urban consumers strive to protect and enhance the purchasing power of their incomes. They demand higher wages, and they amplify their demands in the face of rising prices. Because of the depth of their ties with urban industry, governments resist urban wage demands. But they join in the demands for lower food prices.

Starkly rendered, this is the essence of the political origin of African food policy. But there are other features of the situation which elude so sparse a rendering. Among the most important are elite interests. *Where the elite engages in the production of a food item, policies are not employed to depress its price.* In the case of rice in Ghana, for example, major rice farms are owned by high-level public servants, with the result that rice is sold at domestic prices that lie well above the world market prices, and urban consumers suffer accordingly (see Stryker). Moreover, while the general pattern of protection may be designed to favor the consumer, the actual implementation of protective measures may redistribute income from consumers to elite-level officials. In Kenya, for example, the government forbade the export of certain food items in order to maintain low domestic prices. Nonetheless, in practice, the world price often prevailed. These items simply disappeared from local markets, and it soon became obvious that the administrators of the crop authorities, in cahoots with officials of the border guards, were smuggling the items into the more lucrative world market (for example, see *Weekly Review*, February 9, 1979).

The power of elites and their impact on rural economic and social relations is a theme that we shall stress in later chapters. By examining it, we will move beyond an analysis based purely on the clash of interests between town and country in Africa. We will try instead to show how government policy favors certain rural interests over others, and thereby recruits important allies in the rural sector. Government programs, we will argue, create and nurture rural clients, particularly among elite farmers, and thereby encourage patterns of collaboration that bridge the gap between town and country in Africa.

CHAPTER 3

The Food Sector:

The Use of
Nonprice Strategies

The desire to promote the fortunes of industry and the need to appease the urban areas have led governments to adopt policies intended to provide low-priced food. As has been shown, however, the regulation of internal markets is difficult to achieve. Moreover, the importation of foreign supplies to depress local prices has become an unattractive option. Rising oil prices and demands from industry for imports of capital, machinery, and skilled manpower have intensified demands for foreign exchange. And given the overwhelmingly agricultural make-up of their countries, African governments have responded by promoting programs to reduce food imports by increasing domestic farm production.

This chapter focuses on the production strategies of African governments. It documents their efforts to directly engage in food production and to secure greater private production by subsidizing the costs of farm inputs. One important effect of these strategies, it argues, is their impact on the social and economic structure of the countryside: they confer benefits on the few and promote the fortunes of a small number of privileged farmers. A major reason for the use of these strategies is that they are politically fruitful. Their political attractions will be analyzed in Part Two.

African governments seek to promote food production by means other than raising commodity prices. Many directly engage in agricultural production, using the public treasury to offset production costs and thereby providing cheap food for the urban market. In effect, they enter the market for food and set themselves up as rivals to the peasant producers.

An example is the system of state farms in Ghana. Begun in 1962, the program expanded rapidly; by 1966, there were 135 state farms with a total of 20,800 workers. Hundreds of tractors were imported for these farms; one tractor was provided for every sixty to seventy acres. Between 1962 and 1966, the state farms received approximately 90 percent of the total agricultural development budget for the nation of Ghana (Nyanteng 1978, p. 4; Hill; Gordon).

The state farms were constrained to sell their products below the prevailing market prices. Dadson, for example, compared the prices offered by the state farms with the free-market prices for a variety of products—eggs, poultry, meat, maize, rice, vegetables, and others—and found that the state-farm prices "were consistently and significantly below the free market price" (p. 175). This, of course, was precisely their purpose.[1] One result was that state farms could not meet the demand for their products. The consequences are well illustrated by the attempts in 1964 of the Workers Brigade, which operated a portion of the state farms in Ghana, to market *kenkey*, a popular food item. As recounted by Dadson: "In order to reduce the rising cost of food . . . in the urban areas, the Brigade embarked on a scheme whereby it sold to the public the popular corn food, *kenkey*, at about half the market price. . . . The scheme was popular and successful in Accra, but only for a short time; for, in order to keep the project going, the Brigade had to purchase corn from the local market at prevailing prices for processing and resale" (p. 176). This points to another result of the low-price policy: overwhelming economic losses. Being unable to produce sufficient maize to meet the demand at the controlled prices,

---

1. As Nkrumah had stated in parliament in justifying his production plans: "We must produce food so cheaply that even the worker earning the minimum wage . . . can be fully fed for not more than 2s [shillings] a day" (cited in Dadson, p. 26).

the *kenkey* scheme had to buy maize elsewhere at the market clearing price. As a consequence it soon went bankrupt.

The fate of the *kenkey* project finds its parallel in the economic fate of the overall program of state farms. In a study of the Food Production Corporation farms in the Eastern Region of Ghana in 1971, it was noted that in seven out of eight farms examined, the *annual* gross receipts failed to cover *one month's* bill for wages and salaries (USAID 1975, p. 80)! The Agricultural Development Corporation, which managed most of the farms, accumulated a loss of $4 million in 1964, $7 million by 1965, and over $9 million by 1966 (Miracle and Seidman, p. 43).

The state farms of Ghana thus consumed an enormous portion of the public resources available for agriculture, and they accumulated large debts. In this respect, their fate parallels that of other public production schemes in Africa. The Farm Settlement Scheme of Western Nigeria, for example, consumed £6.4 million over a ten-year period. It has been estimated that over 50 percent of the total capital expenditure on agriculture in the 1962–1968 development plan went into these projects (Nigerian Economic Society, p. 142; see also Hill; Roider). By any criterion, these schemes failed. Investigations revealed that they produced little; what little they did produce, they produced at exorbitant costs; and what they earned was not enough to pay off their initial financing.

The farm projects of Western Nigeria and Ghana used conventional "rain-fed" technologies, but in recent years African governments have increasingly taken recourse to irrigation techniques. One example is the Chad Lake Basin Development Authority, which by 1978 had tens of thousands of hectares under food crop production. The costs of the Chad Basin project are enormous. In 1977–1978, for example, over ₦39 million (*naira*) was budgeted for the River Authority (IBRD 1978). But these costs are simply not being recovered. Commenting in 1978, a World Bank report noted that "the value of the production obtained is less than the operating costs on some of the irrigated land" (IBRD 1978, p. 28; see also *African Business*, April 1980). Even such a famous project as the Gezira scheme in the Sudan, which produces food crops such as sorghum, rice, wheat, and millet as well as cash crops such as cotton, has tended to run at a loss; figures indicate that in not one year between 1971 and 1976 did the Gezira scheme turn a profit (World Bank, *Economic Memorandum on Sudan*, September 27, 1976, Table 4.5).

Although socially costly, both the farm schemes and the irrigation projects tend to be privately profitable for those fortunate enough to gain access to them. Roider, for example, notes that the earnings of those on the Farm Settlement projects of Western Nigeria exceeded those of nearby small-scale farmers; in fact, their earnings approximated those of low-level members of the civil service (p. 105). In the Sudan, farmers in districts with a high density of irrigation facilities earn three to five times the annual revenues of persons located in areas lacking these facilities (ILO 1975e).[2] And data from Kenya show families in irrigation projects earning annual incomes in excess of 20 percent higher than those operating small-scale farms, 200 percent higher than those engaged in pastoralism, and nearly 100 percent higher than those earned by unskilled workers in urban areas (ILO 1978). The private profitability of such schemes is also indicated by the pressures exerted to gain access to them. Interviews with FAO project managers who were supervising irrigation schemes in Ghana disclosed the enormous pressures to which they were subject in the allocation of irrigation plots (Au-

2. Barnett, in his study of tenants in the Gezira scheme (1977), simply fails to take these data into account; it is clear that the tenants on the scheme are in many respects an economically advantaged group in the economy of rural Sudan. It should be stressed that the tenants on government schemes often secure a relatively high level of profits in spite of, and not because of, the way in which the project authorities manage farm production. In Gezirà, the management requires the production of cotton. The tenants contend that they cannot make a profit from cotton at the prices paid for the crop and charged for inputs and services. While their claims may be exaggerated, it is certainly true that farmers can earn more by producing crops other than cotton. As a consequence, they have shifted out of cotton production and into the production of other commodities. The result has been clashes between the government, which earns much of its income from the export of cotton, and the tenants, who resent the loss of income which cotton production entails. Recent reforms, in which the government increased the tenants' share of cotton earnings, have failed to rectify the problem (see discussion in *African Business*, April 1980).

gust 1978). Dadson and Roider each document similar demands for access to position in the state farming projects. By comparison with many other farming opportunities, the state-sponsored schemes promise high private returns. Public food-production schemes thus confer benefits on the fortunate few who gain access to them. The land used on state farms is often seized from small-scale farming communities without compensation (Dadson). The water used by irrigation agencies is often taken from the sources used by small-scale farmers, whether for food production, the dry-season grazing of cattle, or fishing (Scudder 1980, forthcoming). In addition, scarce public services—technical advisors, marketing services, schools, clinics, and extension agencies—that could have been offered to the small-scale farmer are instead put into the service of government schemes. Government-sponsored production units thus often promote the fortunes of a few privileged farmers at the expense of the small farmer in Africa.

Although they consume a significant proportion of the public agricultural budget, these projects nonetheless supply a small fraction of the total market. In the case of Ghana, for example, they provided less than 2 percent of the total marketed output of most commodities (Dadson). In light of such figures, it is inconceivable that they could have much impact on the prevailing level of food prices. Rather, their importance lies in the impact they have made on the social structure of the African countryside.

THE SUBSIDIZATION OF INPUTS

In their efforts to induce increased food production without taking recourse to increased food prices, governments in Africa frequently manipulate the prices of farm inputs. By lowering the price of inputs, they lower the costs of farming; they seek thereby to render farming more profitable, and to attract greater resources to it and evoke greater output from it. What is critical about the means governments employ is that they tend to promote the emergence of coteries of privileged, "modern" farmers. In part, this consequence is intended; the structure of subsidies is designed to promote the adoption of new technologies. But in part it is a byproduct of the way in which the policy is designed and implemented.

The Pattern of Subsidies

Governments in Africa subsidize fertilizers, seeds, mechanical equipment, and credit. They also take measures to promote the acquisition of land for commercial farming.

As illustrated in Figure 1, African governments confer subsidies on fertilizers which run from 30 to 80 percent in value. In many nations, fertilizer is imported duty free. Public support is also given for the purchase of mechanical equipment. In Ghana and Nigeria, farm equipment is exempt from duty; the overvaluation of the ex-

Figure 1.

Levels of Subsidization of Fertilizer for Various African Nations

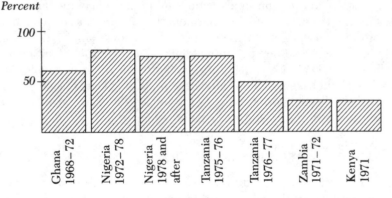

Sources.   *Ghana*: J. Dirck Stryker. "Ghana Agriculture." Paper prepared for the West African Regional Project. Mimeographed. 1975.

Nigeria: International Bank for Reconstruction and Development. "Nigeria: An Informal Survey." Mimeographed. 1978.

Tanzania: Ministry of Agriculture. *Price Policy Recommendations for the 1978–1979 Agricultural Price Review*, Annex 1. Mimeographed. 1977.

Zambia: Doris Jansen Dodge. *Agricultural Policy and Performance in Zambia.* Berkeley, California: Institute of International Studies, 1977.

Kenya: Report of the Working Party on Agricultural Inputs. 1971.

Table 4
Fertilizer Imports, Nigeria

Year	Import value (N-million)	Import quantity (1000 MT)
1970	1.6	34.1
1971	1.8	52.0
1972	4.0	83.0
1973	3.1	84.4
1974	6.1	83.7
1975	12.3	150.9
1976	20.4	207.8

Source. International Bank for Reconstruction and Development. *Nigeria: An Informal Survey.* Lagos: Typescript, 1978, Table 12.

change rate further lowers the perceived price of farm machinery imported from abroad. In Ghana, the Ministry of Agriculture subsidizes tractor-hire services up to 50 percent of actual costs (Stryker; Kline et al.); similar subsidies are provided in Nigeria (Okali). Most nations extend favorable tax allowances to the purchase of farm equipment. Tax holidays are offered to those making major investments in food production or processing; interest payments can be deducted; and favorable forms of capital depreciation are allowed. In Nigeria, an additional capital allowance of 10 percent is offered for expenditures on plant or equipment used in agricultural enterprises. Similar provisions are allowed in Kenya (Kenya 1971; see also Ekhomu; IBRD 1978b; USAID 1976; and Okali 1978).

Data from Nigeria suggest the effect of these provisions. Helped by the influx of revenues from oil exports, duties on fertilizer were canceled and prices subsidized beginning in 1972. In 1975, the duty on mechanical equipment was canceled and subsidies conferred for tractor-hire services and capital credits on the purchase of agricultural machinery. In light of these facts, the data in Tables 4 and 5 are suggestive.

As part of their policies to promote food production, governments also provide subsidies for the development and distribution of improved seeds. In Ghana, for example, the government paid for

Table 5
Tractor Imports, Nigeria

Year	Tractors: tracked		Tractors: wheeled < 40 hp		Tractors: wheeled > 40 hp		Total value farm machinery
	Number	value (N million)	Number	value (N million)	Number	value (N million)	(N million)
1973	202	3.0	397	1.3	468	1.4	6.1
1974	241	2.7	319	1.5	319	0.9	10.8
1975	1209	26.3	2576	13.8	1196	5.1	46.7
1976ᵃ	1922	29.3	1894	7.7	270	2.7	42.9

Source: International Bank for Reconstruction and Development. *Nigeria: An Informal Survey.* Lagos: Typescript, 1978, Table 16.
ᵃJanuary–November only.

one-third of the costs of new maize seeds and three-quarters of the costs of new rice seeds. In Nigeria, the government helped to finance the development of a new, if ultimately ill-fated, variety of maize (see Chapter Two). In Kenya and Zambia the costs of developing and distributing new seeds have been subsidized by the government (Gerhart; Dodge).[3]

To promote the purchase of these new inputs, African governments manipulate the price of capital. In Nigeria, the government has made credit available to farmers at 5 percent below the market rate of interest. In Ghana, the government funded the Agricultural Loan Bank; operating under government regulations, the bank could charge only 6 percent for its loans. The poor recovery rate

3. It should be noted that increased yields from the new varieties of seeds depend upon the use of fertilizers—a fact with important consequences. In assessing its needs for harbor and transport capacity to import sufficient fertilizer for distribution in conjunction with its newly developed maize seeds, the International Institute for Tropical Agriculture wrote: "By 1981, it will require more than three trains per week of over 50 rail wagons (30 tons) each to move fertilizers . . . from the port—if they are purchased in the most concentrated dry form available. Continued use of low analysis materials . . . will more than double the requirement for engines and rolling stock" (IITA, p. 67).

of this bank—63 percent in 1974—further emphasized the concessional nature of the credit offered to investors in food production (see USAID 1976; Girdner and Olorunsola). Lastly, governments have encouraged commercial lenders to move into agriculture by guaranteeing agricultural loans, thereby absorbing some of the risks of these investments.

Governments in Africa have also sought to cheapen the price of land. In Nigeria, the land decree of March 1978 reserves to the state rural lands not under active exploitation. The origins of the decree apparently lie in the desire of the Federal Government to acquire large areas of land "to be leased out on uniform terms to farmers as in the case of industrial estates, on which it 'will be much easier to provide extension services, agricultural inputs, etc'" (from *Guidelines for the Third National Development Plan, 1975–80*, quoted in Gavin Williams, p. 49). Already negotiations are underway in Nigeria between the National Grains Production Corporation and private groups to engage in joint productive ventures on 19 farms of 4,000 hectares each (*New African*, June 1979, p. 97). The effect of the 1978 land decree thus appears to be to move land into commercial production, presumably at a price below that prevailing in the land market prior to the legislation.

In the Sudan, not only government corporations seeking land but also private investors seeking to engage in mechanized farming can obtain land at subsidized prices from the government. By 1968, the government had allocated 1.8 million *feddans* to private individuals (ILO 1975c, p. 1). In many cases, the government used its legal powers to transfer land from traditional production activities, such as nomadic herding, to the mechanized production of food crops without paying, or requiring that the private investors pay, compensation for the loss of rights to use the land for traditional purposes. The effect once again was to place a subsidized price on this input.

Under the terms of the Land Consolidation and Land Adjudication Acts of 1968, the government of Kenya has sponsored the wholesale transferal of land from a jurisdiction governed by customary rights to one governed by private rights. The intention was not to alter the price of land but rather to institute a method of allocating land rights—a private market—that would enhance the efficient use of resources (see Okoth-Ogendo). In practice, however, the reform of land rights has been exploited by those seeking to secure land below the free-market prices. The process by which public agencies have been used to manipulate the land market has been comprehensively documented by Njonjo.

It should be stressed that in reforming land laws, governments in Africa are responding to pressures from potential investors. One of the best examples is provided by Ghana, where potential investors from the southern and coastal communities lobby vigorously for legal reforms in the grain-producing areas of the savannah. The most visible arena for such lobbying is the law reform commission—a commission convened by the government of Ghana to revise codes and statutes, and dominated by lawyers drawn from the more affluent southern portions of the country. In 1977, the government convened the commission to review land law in Ghana; it came forth with a scathing criticism of the prices charged by "landlords" in the savannah. The commission noted that these prices could "become a hindrance for agriculture," and that the needed reforms should include "fixing a reasonable amount of money which should cover customary [obligations]" (quoted in Nyanteng 1978, p. 28). When the landowners were withholding land from the market—in other words, when potential investors could not secure land at a price they were willing to offer—then, the lawyers recommended, the "state should have the power to step in [and] make grants of vacant lands in that area" (ibid., p. 29). These recommendations constitute a plea for changing land law so that the state would have the power to depress the price of land for the benefit of private individuals who seek to invest in farming.

IMPLICATIONS FOR THE COUNTRYSIDE

The governments of Africa thus intervene in the markets for farm inputs—fertilizer, farm machinery, seeds, credit, and land. They do so in order to depress the price of the inputs and thereby enhance the profitability of farming. It is difficult to assess the impact of these programs on aggregate output or on the cost of food. It is easier to assess their impact on the distribution of income in the countryside. It is commonly and almost universally found that the poorer, small-scale, village-level farmers do *not* secure farm inputs that have been publicly provisioned and publicly subsidized as part of programs of agricultural development. The evidence suggests that the benefits of these programs have been consumed chiefly by the larger farmers, sometimes at the expense of their smaller counterparts.

Indirect Evidence

The best support for this contention is contained in investigations into the failure of small-scale farmers to adopt new technologies. Time and again these investigations reveal that conventional explanations are wrong. The village-level farmers do in fact know about the advantages of new seeds and of fertilizers; they do want to use them; and they are especially interested in securing them at their publicly supported prices. The reason for the failure of the new technologies to "diffuse" through the rural community thus has little to do with the attitudes of the village farmers themselves, as is commonly claimed. The problem instead is that the inputs are often not available.

One Ghanaian study of the failure of small farmers to adopt chemical inputs noted that "even though the farmers are prepared to purchase and use . . . fertilizer to improve their yields, fertilizer and chemicals were largely unavailable to them" (Armah, p. 20). In reviewing similar studies in Nigeria, the World Bank noted that "numerous micro-studies have been conducted in recent years indicating that [only] about 10 percent of the farmers do not understand the value of fertilizer or feel it will not produce yield responses. . . . All of the numerous studies identify the primary limiting factor as fertilizer unavailability" (IBRD 1978b, p. 34).[4]

4. There are, of course, many other reasons for the failure of fertilizer programs in Africa. Even when fertilizers are available to the small-scale farmers, they are often not available at the right time. Moreover, Africa contains a great diversity of soils, and little research has been conducted on which fertilizers are appropriate for which soils. This lack of knowledge leads to the distribution of inappropriate vari-

Similar results have been found in studies of government-sponsored credit programs. Investigations in Ghana reveal a strong demand for public credit on the part of small-scale farmers; they also reveal an enormous frustration with the nonavailability of loans and an impressive expenditure of energies in attempts to extract them from the governmental bureaucracy (see Armah). A review of local-level studies in Nigeria suggests a similar pattern (IBRD 1978b, pp. 35–36).

Although governments have sought to increase the production of food by supplying farm inputs at subsidized prices, the experience of small-scale farmers has been that these inputs remain scarce. But the government programs have been welcomed enthusiastically by wealthier and more powerful people. The resources allocated through these programs have been channeled to those whose support is politically useful or economically rewarding to the state—that is, to members of the elite.

Direct Evidence

Perhaps the best evidence of these trends comes from the savannah regions of West Africa. In response to government efforts to promote the supply of inexpensive food for the cities, there has arisen a cadre of commercially oriented, mechanized farmers—a group whose existence is predicated on the provision of government subsidies and whose membership consists largely of wealthy and politically influential members of the urban elite. An example would be the mechanized farmers of northern Ghana.

Mechanized farming began in the northern regions of Ghana in the 1960s, but burgeoned in response to the incentives provided in the late 1960s and early 1970s to encourage domestic food production. Under the policies mounted by the Ghanaian government, the northern farmers, like all farmers in Ghana, qualified for subsidized

---

eties. Moreover, extension agents, when they exist, often are poorly trained and give inappropriate advice. The result is that the farmers obtain few gains from the use of this input, thus weakening the incentives to adopt fertilizer or fertilizer-responsive varieties of crops.

seed, fertilizers, and credit; the evidence suggests that, unlike the small-scale farmers, they actually received these benefits. According to the agricultural census of 1970, the Northern and Upper Regions had only 22 percent of the total agricultural holdings in Ghana and produced less than 20 percent of the total value of Ghana's agricultural output. But one source reveals that over 75 percent of the fertilizer imported into Ghana in 1974, and virtually all of the improved seeds, went to the Northern and Upper Regions (USAID 1975, pp. 137–146). The government vigorously promoted the use of mechanized production techniques by those seeking to invest in the area. As one appraisal noted: "a relatively large number . . . of tractors and associated equipment . . . are available for initial land preparation. . . . The charges are artificially cheap owing to an overvalued exchange rate which keeps capital costs for tractors, equipment, and spare parts down" (ibid., p. 94). By 1968, the government had placed 907 motorized units in the Upper and Northern Regions (Kline et al., p. 388). And the evidence strongly suggests "that the tractor-hire service was well received by progressive farmers who were anxious to make use of it. . . . Apparently, the services offered were economical, from the farmers' point of view" (Kline et al., p. 122).

Evidence of the relative success of the large-scale farmers in securing subsidized credit is that in 1974, 56.3 percent of the total funds loaned by the Agricultural Development Bank were distributed to the 3.5 percent of applicants who were authorized to borrow £20,000 and above (Rothchild 1979). Moreover, government reports document a low level of repayment by the large-scale farmers. Only 44 percent of the agribusiness ventures, the large operations characteristic of this area, were in good standing in their loan repayments in 1974, compared with an overall level of 63 percent for farmers as a whole (USAID 1976, p. 16). Rates of repayment by the large-scale farmers were thus lower than that by other farmers. In particular, they lay below the rate of repayment by the small-scale farmers, who were faced with a harsh government credit policy: loans would be denied to any member of a village cooperative that included a farmer who had yet to repay a government loan.

My interviews with low-ranking members of a credit agency in Ghana furnish persuasive if impressionistic evidence of the role of privilege in securing subsidized credit. Respondents agreed that credit for food crops was not allocated according to commercial criteria but rather according to patterns of friendship and influence. They stressed that their attempts to apply commercial criteria in evaluating applications for funds led to rebuffs by superiors in the organization. Applicants would go over the heads of the professionally minded lower staff, and the staff would subsequently receive directives ordering the release of funds to specified individuals. "Connections" have thus played an important role in structuring the allocation of loan funds to the commercial food crop producers in the savannah areas of Ghana.

Equally striking has been the manipulation of political connections to purchase land in the savannah region. We have already seen that private investors have sought to reform land law in Northern Ghana. The evidence suggests that while awaiting these reforms they have used existing institutions to secure access to farm lands.

In contrast with the rest of Ghana, in the savannah areas of the north the state can exercise direct control over rights to "unused" or "waste" lands; these rights are allocated by the national department of lands. Members of the urban elite who seek to invest in farming and who have connections in the national bureaucracy have used the power of the lands department to secure acreages for food production. Indicative of this are the disputes involving the Karaga people of Dagomba and the Builsa people of the Upper Region on the one hand and the government bureaucracies and commercial farming interests on the other. According to one report:

Both Karaga and Builsa have been involved in disputes over land with stranger farmers—Karaga with Nasia Rice Company, and Builsa with a group of . . . farmers supported by political allies in the regional government. Both areas are latecomers to rice farming, and have learned from the mistakes of other [northern] communities. . . . Both have refused to sanction Lands Department leases. . . .

But these examples are exceptions: they could not be repeated in areas where a significant number of stranger farmers have already made . . . farms. And at least in Karaga and Builsa it would be hard for the traditional authorities to reclaim land from the tenants once it had been leased to them, as some powerful figures in Ghana are among their number. [*West Africa*, April 3, 1978, p. 647]

Using political connections to secure land, publicly subsidized credit and forgiveness of debts, publicly subsidized and allocated fertilizer, and highly favorable terms for the importation and financing of capital equipment, influential members of the urban elite with close ties to the managers of the public bureaucracies have thus entered food production in the northern savannah areas. The result has been a transformation of the pattern of agricultural production in the savannah zones. Rather than small-scale peasant farmers, the new entrants are large-scale commercial producers. Instead of hoes and oxen, they use tractors and combines. A major consequence of government efforts to promote food production in this area has been the development of disparities of wealth, social status, and political power within the savannah region.

When similar policies have been adopted elsewhere in Africa, the consequences have been much the same. One example is the growth of mechanized farming in the Sudan, with its debilitating effects on the environment and the threat it poses to pastoral production. Another is the development of large-scale farming in regions of pre-Revolutionary Ethiopia (see Cohen and Weintraub). A third is in the Rift Valley of Kenya, where government programs have promoted the mechanized production of grains, particularly wheat and barley, in what were formerly grazing areas. The production of these crops is sponsored by state grain corporations headed by persons of enormous political influence. A similar pattern appears to obtain in the middle-belt regions of Nigeria, where state corporations and politically important individuals are investing in mechanized schemes for the production of food. The policy responses of African governments to the problem of urban food supply thus appear to be leading to the entrance into the countryside of politically influential elites—elites who seek to augment their fortunes by engaging in food production, and who adopt farming technologies that fundamentally alter the social and economic patterns of the African countryside.[5] In other areas of the developing world, the existence of elites deriving their wealth and power from agriculture antedates the commitment of national governments to programs of economic development. In these areas the politics of development became in part the politics of displacing these existing elites, as urban interests attempted to secure their capitulation to the new economic order. By contrast, at the time of the commitment to industrial development in much of Africa, the countryside contained few persons of landed power. It is the programs in support of economic development that have promoted the growth of such elites in the rural areas. The initial push toward industrialization has thus encountered far less overt resistance from the rural areas of Africa.

As will be seen in later chapters, however, these privileged farmers, despite the fact that they owe their position to governments dominated by urban interests, soon give voice to producer interests. What the small farmers cannot demand, the elite farmers do.

5. Moreover, the evidence suggests that in reaping disproportionate benefits from public programs, the large farmers do so at the expense of small-scale producers. Certainly the redefinition of land rights and the subsequent reallocation of land between "traditional" and "commercial" sectors represents such a redistribution. So, too, does the evidence concerning subsidized loan programs, already cited. Besides receiving the bulk of the loans from such programs, large farmers also more frequently default on them; the costs are passed on to the small-scale farmers in the form of higher interest rates. Redistribution also takes less obvious forms. In 1976–1977, for example, 50 percent of the cost of the fertilizer subsidy of Tanzania was to be paid for by funds from the crop authorities; the authorities in turn received their funds in the form of deductions from payments to farm producers. Insofar as such deductions are made from payments to both small farmers and large ones, and insofar as the fertilizer tends to be consumed by the larger farmers, the subsidy redistributes resources between two kinds of farmers. In Kenya and Tanzania, the costs of some farm inputs are financed by cooperatives; and studies show that while the costs are born equally by all members in the form of subscription payments, the benefits are consumed disproportionately by the larger members (a review of these studies is contained in Raikes). Public financing of the costs of farming thus leads to patterns of subsidization that favor the larger farmers, and at the expense of their small-scale counterparts. For further documentation of the large-farm bias in the provision of agricultural services, see Leonard, Bottral, Hunt, and Kenya (1971).

At present, governments have successfully co-opted them; they are rural allies of the regimes in power. But the basic conflict of interest remains, and as development proceeds and the community of large farmers expands, they and the interests they represent should become more powerful. Africa will clearly not remain immune to the political conflicts between agrarian and industrial interests that are an inherent part of the development process.

CHAPTER 4

The Emerging Industrial Sector

Thus far we have analyzed government interventions in the markets for products that farmers sell and in the markets for products they use in farming. There remains a last major market to be explored: the market for the commodities that farmers consume, and in particular the goods they purchase from the urban-industrial sector.

Like governments throughout the developing areas, the governments of Africa try to promote industrial development, and every government in Africa has pledged to develop its national economy by creating domestic industries. This chapter will show that a major strategy for promoting industrial development has been to shelter new firms from meaningful economic competition, whether domestic or foreign. Consumers therefore inevitably pay for a part of the cost of industrialization in the form of higher prices. The consumers who concern us here belong to the farming population.

COMMERCIAL POLICY

In some African countries, governments have imposed commercial barriers to foreign competition rather quietly. In Tanzania, for example, the government is officially opposed to the use of public power to promote the economic fortunes of private investors; nonetheless, it does seek the formation of local manufacturing capabilities, and as part of its policy of socialist development, it seeks to promote state-backed industries. The result has been the adoption of a structure of commercial protection that shelters local industries (Rweyemamu; see also Clark).

In other countries restrictions on imports, at least initially, have been imposed more in an effort to conserve foreign exchange than in an effort to promote industrial protection. Nonetheless, the measures rapidly become an instrument of economic protection. In Ghana, for example, significant restrictions on foreign trade were first introduced following large trade deficits in the early 1960s; in response to this crisis, the government imposed import licensing and foreign-exchange controls. As Killick notes, it was not long before criteria for allocating foreign exchange were formalized, and one of the key criteria "by which the import planners were required to allocate licenses was that of *protecting local industries*" (Killick, p. 278).

In other cases, however, the protective content of government policies has been explicit; it has been publicly affirmed in an effort to attract investments. Thus, Kenya in 1959 incorporated a schedule of explicitly labeled protective tariffs into its commercial legislation (Swainson 1977a, p. 149). In pre-independence Nigeria, Oyejide reports, the tariff structure was basically "revenue oriented." Within a year after independence, however, "the protection of the domestic market to encourage industrialization *via* import substitution had become an official policy; and since no serious balance of payments crisis arose until the tail-end of 1967 [with the civil war], it may be assumed that the tariff changes that took place within this period were primarily a direct consequence of this official policy" (Oyejide, p. 58). Commercial protection for domestic industries remains a prominent feature of Nigerian policy, as evidenced by the last major budget speech of the departing military government (see *African Business*, May 1979).

Governments offer tariff and import protection in efforts to attract foreign investment. The most thoroughly documented case is

Kenya, where Langdon has analyzed the negotiations between the New Projects Committee of the Government of Kenya and the representatives of foreign firms. The demands most commonly made in these negotiations were for protection from foreign competition, either through tariff protection or physical restrictions on imports (in 53 percent of the negotiations), and for concessions in tariffs and restrictions on imported supplies and capital equipment (in 32 percent of the negotiations). Over the period 1965–1972 protection was granted to manufactured products in 90 percent of the cases, and concessions were accorded for the necessary inputs in every case considered (Langdon; see also the works of Swainson). In a less detailed analysis, Young notes the adoption of similar measures in Zambia. And though we lack comparable data for other countries, government-offered incentives in the search for foreign investments appear to be standard fare throughout Africa.

Tariffs are one means of protecting local industries. In the African setting, physical restrictions on imports are even more important. Where they are a feature of commercial policy, the committees that control the allocation of licenses to import or permits to use foreign exchange become key centers for the allocation of economic shelters.

The operation of such committees has been briefly described by Fajama for Nigeria, Leith for Ghana, and the ILO-UNDP mission for the Sudan (ILO 1975d). Macrae gives a fuller treatment of the relevant body in Kenya, the Committee for Industrial Protection. He notes that one of the Committee's main tasks is to issue import licenses, and that the procedures it adopts give protection to key domestic industries. The Committee acts in response to petitions. As Macrae stresses:

Certain items are referred to specific bodies before an import license is granted. The Ministry of Agriculture must approve imports of millet and grain sorghum . . . cereals . . . prepared animal feeds, oranges, jams, beans, garlic, frozen vegetables and fertilizers. Import licenses for paints are issued on the recommendation of the Association of Local Manufacturers, as also are motor vehicle batteries; licenses for importing jute and sisal bags and sacks are issued on the approval of the Jute Controller. In most cases this is to confirm whether local supplies are available, in which case license applications are refused. Some importers are granted a monopoly outright, e.g., import licenses for iron and steel-wire are issued to the Kenya Industrial Estates only. [P. 8]

Patterns of Protection

Evidence on the pattern of protection created by African governments, though widely scattered, exhibits one common feature: the level of effective protection exceeds the level of nominal protection. Both forms of protection result from barriers that favor domestic producers. Nominal protection is protection given to the price of products; when governments impose tariffs or quantitative restrictions on imports, they enable domestic prices to rise above the price of foreign goods. Effective protection is protection given to the profits of industries; it takes into account not only the impact of trade barriers on the prices of products but also on the costs of goods used in their manufacture. To encourage the formation of industries, governments must protect not only prices but profits. When they use tariffs and trade barriers to increase the price of a product, they must, if they wish to create incentives for its manufacture, therefore refrain from comparably increasing the prices of goods used in its production. It is indicative of the efforts of African governments to create incentives for the formation of industries that the level of effective protection exceeds the level of nominal protection; few barriers are placed on the importation of goods used by the industries but protection is given to their products.

Governments in Africa have used commercial policies to strengthen incentives for local production. Evidence from the Sudan suggests a pattern of high nominal rates of protection but even higher levels of effective protection. Thus a team from the International Labor Office found, for industry, an "average effective rate of protection of 170 percent for 1971." It went on to comment that "since then further tariff concessions have undoubtedly increased protection," and to note that "this contrasts with the previous estimate of minus 27 percent for agriculture and illustrates the considerable inducement given by price incentives policies to industrial as opposed to agricultural development" (ILO 1975d, p. 35). Rweyemamu, in his study of Tanzania, concludes that "in most industries, the effective protective rates are considerably *greater* than the nominal rates," mainly because "duties on most raw materials and other inputs are either zero or very low" (p. 133). For Kenya, a World Bank study (and the Institute for Development Studies' papers it draws upon) reveals "the classic tariff structure, with average nominal duties falling from 29.6 percent on consumer goods to 18.0 percent on intermediates, and 17.7 percent on capital goods" (IBRD 1975, p. 265)—a pattern that would, of course, produce a rate of effective protection exceeding the rate of nominal protection. A similar pattern is found by Oyejide for post-independence Nigeria (Oyejide, p. 59).

Clearly, then, African governments have erected structures of protection that systematically favor the formation of domestic manufacturing capabilities. What is also suggested is that they have done so *in particular* for industries which produce goods for final consumption. This is suggested in Oyejide's data, where the highest rates of both nominal and effective protection occur for consumer goods. As Oyejide himself concludes, "the bias of the tariff structure [is] clearly in favor of consumer goods" (p. 58). Textiles, bicycles, processed foods and beverages, footware, clothing—these are the kinds of products most favored by the tariffs Nigerian policymakers have imposed. A similar pattern is documented for Tanzania, where Rweyemamu concludes that "there seems to be a tendency for consumer goods industries, and in particular the less durable and luxurious types, to be heavily protected" (p. 133). Included among the specific products protected in this manner are bicycle tires and tubes, sugar, beer, biscuits, soap, clothing, footwear, matches, and tobacco (ibid., p. 134). Similar patterns have been detected for Zambia (Young), the Ivory Coast (IBRD 1978a), and Kenya (IBRD 1975).

Thus, to promote industrial development, African governments construct protective barriers between the world and domestic markets which shelter local industries from foreign competition. And they give particular protection to industries that produce goods for final consumption.

SHELTER FROM DOMESTIC COMPETITION

Public policies to promote domestic manufacturing often inhibit domestic competition as well. In some cases, restrictions on competition at home are a byproduct of measures taken to restrict competition from abroad. In both Ghana and Kenya, for example, the tariff laws are written so that the incidence of protection is designated at the "six-digit" level of industrial classification (Pearson et al., p. 14; Macrae, p. 5); in effect, then, protection is extended to the individual firm. In Kenya, licenses to import goods listed on what is called schedule D, or materials for the manufacture of such goods, may be issued only after the Director of Trade determines that there is "no objection" to this use of foreign exchange. My interviews in Kenya reveal that local firms lobby strenuously to place their products on schedule D. They do so because they can then "object" to imports of their product or of material which could be used for its manufacture. The trade law thus shelters them from domestic as well as foreign competition. Most trade programs involve the allocation of quotas or licenses; these permits to import are often distributed in accordance with historical market shares. Use of this criterion has been recorded for the Sudan (ILO 1975d), Ghana (Pearson et al.), and Nigeria (Fajama). The effect, of course, is to freeze existing patterns of competition, thereby preventing the growth of more efficient and lower-cost firms.

Lastly, bureaucratic procedures for extending protection from foreign competition tend to give an advantage to larger firms, and this too promotes market concentration. Larger and better staffed firms have a systematic advantage in preparing justifications for demands for protection, or for rations of foreign exchange; in devising estimates of costs and in gathering and analyzing supporting data; and in handling the volume of paperwork involved in securing administrative action. As a World Bank study of Kenya found: "The entire system benefits large and well-established firms. Dealing with the bureaucracy requires time and money—both assets of large firms. The more complex the system becomes, the more important are these assets. . . . [Several new] firms have been squeezed out by . . . the allocation of quotas and the costs of dealing with the bureaucracy, [although] others with good connections have obtained licenses" (IBRD 1975, p. 298).

The restriction of competition in the domestic economy is not merely an unintended consequence of the procedures used to govern relations with the international market, however. The consolidation of industries is sometimes done on purpose. As Leith noted for Ghana: "The import-license system, since it had virtual life and death powers over most industries, came to be used as an industrial licensing system as well. The Ministry of Industries saw a conflict between the need for competition among domestic producers and the wasteful expenditure involved in duplicating underutilized domestic facilities, but generally resolved it . . . in favor of 'rationalization' of industries and against new entrants" (Leith, p. 32).

In other instances, the rights to import capital goods and inputs necessary for manufacturing a particular product have been purposefully restricted to particular enterprises. To secure the erection of an automotive assembly plant, the government of Kenya gave British Leyland the sole right to import particular parts and machinery (Swainson 1977a), p. 305); similar privileges were extended to Firestone to secure its investment in a domestic tire plant (Langdon, p. 172). The effect was the promotion of a virtual monopoly for both firms in their respective industries.[1] The extension of exclusive rights to import has been used to promote investments in Zambia as well. There, too, it has resulted in the creation of domestic monopolies in several industries: cement, food processing, matches, sugar, building materials, petroleum, and textiles being cases in point (Young, pp. 193ff).

Governments thus use commercial policy instruments to promote the formation of their nation's industrial and manufacturing capabilities; and in so doing they often restrict not only foreign competition but also competition within the domestic market. It should also be noted that other policies have promoted industrial con-

---

1. Firestone's "concession" was limited to ten years. In 1979 the ten-year period came to an end, and a second firm now proposes to enter the Kenyan market (*Weekly Review*, January 26, 1979; also *African Business*, May 1979). Firestone is retaliating by increasing its production, thereby making entry less attractive.

centration: among these are tax credits, accelerated depreciation allowances, subsidized interest rates, and preferential duties on capital equipment. All these have been used by governments to promote the importation of capital and thereby lay the foundations for industrial development. Moreover, in negotiations with foreign investors, governments tend to favor those who promise larger investments. The result has been the adoption of capital-intensive technologies which are most efficient at high levels of output. But, by and large, the domestic markets of the African countries are small; there are few people and they are poor. Given the capital-intensive nature of the new firms and the small domestic markets, there tends to be idle capacity in many industries, and the incentives are thus strong to secure a reduction in the number of firms.

Again, though the evidence for this assertion is scattered, it tends to be persuasive. In a survey of forty-four Kenyan industries, for example, the World Bank noted that in only twelve of them was there a "reasonably full utilization" of productive capacity (cited by Godfrey and Langdon, p. 115). In Ghana, government estimates suggest that for state enterprises, output was 29 percent of capacity in 1963–1964. In 1966, actual manufacturing output was one-fifth of the single-shift capacity of installed plant, and in 1967–1968, manufacturing firms in Ghana used only 35 percent of their estimated capacity (Killick, pp. 171, 196).

It is clear that this idle capacity is perceived as excess capacity. A 1969 survey of the managers of manufacturing firms in Ghana revealed that "Only 24 percent of them thought that the market was big enough to absorb the full capacity output of the industry at ruling prices and 63 percent believed that industrial capacity exceeded the market size at any feasible price. No less striking, 37 percent of the respondents thought their own capacity exceeded the market" (Killick, pp. 197–198).

Such beliefs furnish incentives to restrict competition. Evidence suggests, for example, that Dunlop chose not to enter the East African market for bicycle tires because Avon Rubber and Bata already had capacity "well in excess of the level of domestic demand" (Eglin, p. 117); this left two major firms in the industry. In a well-documented case, Swainson indicates how firms in the Kenyan cement industry repeatedly merged until only two companies remained; these firms then negotiated a division of the market, one producing 80 percent of its output for export and the other 90 percent of its output for internal use (1977a, p. 193). And with this agreement there came a major rise in price (Eglin, p. 119). The East African paint industry was similarly characterized by initial overexpansion and vigorous price competition. Eventually, the four remaining firms agreed to form a cartel, called the East African Paint Industries Association. This cartel then secured tariff protection to restrict foreign price competition while implementing an internal price agreement within the East African market (Eglin).

An even more recent example comes from the Kenyan textile industry. In the early 1970s, Lonrho, the West German Development Corporation, and local Kenyan investors financed construction of the Nanyuki Textile Mills. In December 1977, the venture failed. An investment of £8 million and the jobs of 750 workers had been imperiled by the inability of the mill to produce cloth at competitive prices; as the management contended, "the Kenyan market was saturated" (*African Business*, September 1978, p. 31). Recently the firm has been reopened, under arrangements that are instructive. Its assets were purchased by a competitor, Mount Kenya Textile Mills, and the reopening of the failed firm was made conditional on a government guarantee banning the importation for sale in Kenya of secondhand clothing (*African Business*, December 1978, p. 60). Here, as elsewhere, internal and external competition has been restricted in order to promote the formation of domestic industries.[2]

INDUSTRIAL STRUCTURE

We lack good data on the structure of the industrial sector that has emerged as a result of these policies. But what little we do have tends to suggest that the total number of firms is small; that in each industry there are few firms; and that within each industry produc-

2. Similar steps were taken to safeguard Kafue Textiles in Zambia (see Young, p. 194).

tion tends to be concentrated within a very small proportion of establishments.

Materials from Kenya illustrate these points. In the manufacturing sector in 1972, there was a total of only 3,687 establishments. To appraise this figure meaningfully, at least two adjustments must be made. One is to adjust for the very small, highly specialized fabricators, such as local tailoring and carpentry shops. This reduces the number of firms by 1,715, leaving 1,972. The other is to look at the number of establishments in particular industries. We then see that the number of units producing sugar is eight; the number slaughtering and dressing meat, eight; the number ginning cotton, ten; the number spinning cloth, nineteen; the number manufacturing textiles, two; and so on (Kenya 1977, pp. 95ff). Thus not only are there few manufacturing establishments in Kenya, but also in any particular sector the number of establishments is small.

Elsewhere we find a similar pattern. In Tanzania, state or state-associated firms controlled 57 percent of the manufacturing sector in terms of value added, or 47 percent when measured in terms of employment (Clark, pp. 64, 126). As Clark notes:

The parastatal [state-associated] sector is characterized by a heavy dominance of a few firms. The government has not created a sector composed of medium-size operations but one in which a few firms own most of the assets. . . . In many sectors only a few firms dominate. In mining, construction, and electricity, one firm has over 80 percent of the assets in each sector, and in agriculture and transport two firms have over 80 percent of the assets in their respective sectors. . . . Nine manufacturing firms (21 percent of total) own 74 percent of the assets in the sector. [Pp. 118–119]

In Zambia in 1969, there were but 431 manufacturing establishments (Zambia, 1971). Again, the number of firms per actual industry was small: two leather and footwear establishments, three spinning establishments, three firms producing vegetable oils, three producing canned goods, and so on (*ibid.*). And as in the case of Tanzania, the state-associated firms, organized under the Industrial Development Corporation (Indeco), controlled in excess of 50 percent of the manufacturing sector and in many instances operated virtual monopolies. As Young states: "For many of the Indeco companies, and indeed for many private ones, the business environment was often less than ruthlessly competitive. Because of the scale of their operations, the more important new industrial projects were generally in a monopolistic position in the domestic market" (p. 203).

Ghana, in 1969, had 356 manufacturing firms (Ghana, Central Bureau of Statistics, 1971). In keeping with the state-centered thrust of its industrializing strategy, public enterprises dominated many basic industries. And as Killick notes: "Many of Ghana's state enterprises were monopolies or were selling in highly imperfect markets. Industrial statistics indicate that, in 1969, 83 percent of the total gross output of state enterprises was produced in industries in which state concerns contributed 75 percent or more of the total output of the industry. In six industries state enterprises accounted for the whole output" (pp. 220–221).

Obviously, these data leave much to be desired. They nonetheless suggest that the policies designed to promote industrial formation in Africa have produced a highly concentrated industrial structure. The total number of firms is small. Moreover, within particular industries, there exist few firms and a small number appear to produce a high proportion of the total output.

Consequences

In this chapter we have explored some of the basic features of policies affecting the growth of the industrial and manufacturing sector in Africa. These policies shelter firms not only from foreign but also from domestic competition. One result is that many inefficient firms survive in the African market.

Evidence of this is contained in the figures on excess industrial capacity, which suggest that many firms fail to operate at the cost-minimizing levels of output. Further evidence is contained in qualitative descriptions of the difficulties of operating modern plants under conditions prevailing in Africa. Schatz (1977), for example, in describing the problems bedeviling new enterprises in Nigeria, reports that equipment was ordered at a long distance from its place of design and manufacture; the result was economic losses from inappropriate equipment and from delays while awaiting corrections in deliveries. Because of long distances and problems in transporting, offloading, and storing, machinery often arrived in poor condition, and this led to further losses. Once they arrived, the machines were often improperly installed; the results were either high operating costs or costly delays while awaiting rectification. Often the equipment could not employ local inputs. A furnace might be unable to work local silicons, or a textile plant might be unable to secure fibers of appropriate length from local producers. Problems such as these, Schatz notes, repeatedly plagued efforts to establish new firms. Killick paints a similar picture of the problems facing firms in Ghana. The obvious corollary of their discussions is that the firms are inefficient and incur high costs, and that without substantial protection from meaningful economic competition, many of them could not survive.

The survival of such firms entails substantial costs, and it is consumers who pay.[3] When protection is offered against lower-cost for-

3. Thus Nkrumah is quoted as stating: "It may be true in some instances, that our local products cost more, though by no means all of them, and then only in the initial period. . . . It is precisely because we were, under colonialism, made the dumping ground of other countries' manufactures and the providers merely of primary products, that we remained backward; and if we were to refrain from building, say, a soap factory simply because we might have to raise the price of soap to the community, we should be doing a disservice to the country" (quoted in Killick, p. 185). Nonetheless, it is also true that the public as a whole bears the costs of government policies which reward particular private interests. Many therefore take a different view, based on a clear perception of the redistributional nature of the policies designed to promote local manufacturing. Such a view is expressed in the following letter penned by one of Nkrumah's countrymen. The conflict in viewpoint is sharp and fundamental, though raised in droll language: "In Ghana, if a company is able to produce an inferior type of product which has been lying in a warehouse unpatronized for years, it then runs to the government claiming that . . . the government should stop the importation of such items. This is usually quickly agreed upon . . . then all of a sudden, the papers tell us that such and such a product is being banned forthwith since we are self-sufficient in that field. . . . Because of Union Carbide, the importation of batteries was restricted and a torchlight battery sells at between ₵ 2.50 and ₵ 3.00; because of G. T. P. and Akosombo Textiles, no importation of cover cloths, and a piece of Dumas sells at between ₵ 150–₵ 200; because of Lever Broth-

eign goods, the result is an increase in domestic prices. And when domestic competition is restricted, firms can secure prices that give them higher profit margins (for evidence, see House, p. 12).[4] The result in both cases is a rise in consumer prices.

DISCUSSION

In earlier chapters, we have argued that pressures from the urban sector generate demands for policies to secure lower consumer prices. In this chapter, we have stressed the role of urban interests in securing policies that increase prices to consumers. The contrast is significant and important; and the apparent conflict can be resolved in a way that gives insight into the interplay of economic interests in the policy-making process.

We can begin with a single industry. It is reasonable for those who derive their incomes from the production of a product to seek a higher price for it. This is true of workers as well as the owners of firms, for both derive their incomes from the production of a particular good. But they spend their incomes widely, devoting but a small fraction, in most cases, to the purchase of the good they produce. Thus they benefit from an increase in its price.

Insofar as governments respond more readily to business combinations than to individuals, it is also reasonable for those who derive their incomes from making a particular product to combine with persons from other industries in seeking protection for their products. Makers of tires, for example, can often do better in seeking government support for higher prices if they receive at least the tacit backing of the makers of bicycle frames. And it is advantageous for persons from several industries to combine in this manner.

ers (Ghana) Ltd., you can't import any type of soap, all you can get (toilet soap) ranges from ₵ 2–₵ 2.50. . . . Yet all these factory managers claim they can meet the demands of the entire population" (*West Africa*, October 16, 1978).

4. In light of what we have noted above, it is instructive that House was unable to disentangle two separate effects: one arising from industrial concentration and the other from capital requirements to start new firms. Plant size and economic concentration went together, and both related to the capacity of firms to secure favorable price-cost margins.

Those who derive their incomes from the production of tires would gain from the increase in earnings which a rise in their own price entails; and they would lose only a portion of this increase from having to pay higher prices for bicycles and flashlights, for example, with whose producers they may have combined in their lobbying efforts.

There is a limit to this logic, however. Not all industries are equally attractive partners in this price-setting game. In particular, if one industry's product requires the expenditure of a very high portion of a person's budget, then persons will look for other industries when seeking partners with whom to combine in petitions for higher prices. In Africa, as in other poor areas, food is such a product; as much as 60 percent of the average urban dweller's budget is spent on food purchases (Kaneda and Johnston). In the formation of combinations to secure price increases, food producers are therefore unattractive partners, and tend to be excluded from price-setting coalitions (see Bates and Rogerson). Demands for higher prices for industrial products and lower prices for agricultural goods are thus an expected result of the free interplay of interests in attempts to lobby and thereby influence product prices.[5]

Other factors also help to resolve the apparent contradictory behavior of urban interests. By offering high levels of effective protection to an industry, the government can secure higher returns to all factors operating in that industry; this provides an incentive for capital to move into that industry, but it also enhances the value of labor. Labor and capital can both share in the gains generated by protection. The demands of labor which we discussed in Chapter Two are thus, ironically, assuaged by policies that try to provide incentives for capital investment by conferring higher prices on manufactured products (see also Arrighi).

It should be noted that not all farmers suffer as a consequence of this dynamic; certainly, they do not suffer equally. As noted in Chapter Three, for large and privileged farmers, the impact of adverse prices is offset by the conferral of subsidies. Moreover, the

5. Relevant here are the analyses of the terms of trade between agriculture and industry. See the works of Maimbo and Fry, Dodge, Sharpley, and Killick.

producers of some crops are able to secure increases in prices for the goods they sell, and these help to offset the higher costs of the goods they buy. In particular, those who, like the food producers, are able to avoid government marketing channels can shelter themselves from the adverse shift in prices. The small farmers and the farmers who produce crops whose marketing is effectively dominated by government marketing agencies are less able to avoid government policies and so suffer most.

CONCLUSIONS

As with governments elsewhere in the developing world, governments in Africa seek to industrialize. They do so in part by sheltering domestic industry from foreign competition. They also protect firms from domestic competition. Characteristically, industries in Africa are dominated by a few large firms; sometimes they are dominated by a monopoly; and often, the major firms are government-owned. Under such sheltered conditions, inefficient firms survive. And consumers, including farmers, pay higher prices.

Many would argue that the burden of higher prices represents a cost of the transition to an industrialized economy. Bergsman, for example, reappraised the economic growth of Korea, the Republic of China, Brazil, Singapore, and other semi-industrialized countries and stressed that their development involved passage through an initial stage that closely resembles that characteristic of contemporary Africa. Nonetheless, while these conditions may be a necessary prelude to later industrialization, they clearly are not a sufficient condition for it. This argument is supported by Bergsman's analysis, which notes the failure of other economies, and it should give pause to those who would see in the experience of these countries a promise of successful industrialization in Africa.

Several characteristics distinguish the now semi-industrial states from their less successful counterparts. One, Bergsman contends, is their policies toward agriculture. In addition to the protected conditions afforded their industries, many of the governments of these states also provided a strong stimulus to farm production: "favorable prices plus heavy investment plus good access to inputs," in

Bergsman's words (p. 80). Such policies contrast sharply with those found in most of Africa. Another distinctive characteristic of successful cases is the existence of large markets for manufactured products. Either because of their exceptional size (as in the case of Brazil) or because they specialized in the manufacture of exports (as in the case of Korea, Hong Kong, or Singapore), the successful countries tended to have access to larger markets. In the first case, they had little incentive to maintain few firms; in the second, they lacked the power to exclude competitors. Large markets therefore promoted conditions under which efficient operations became an established part of the economic order.

In Africa, few nations attempt to export manufactured products. Most have small populations and the majority of their citizens are poor. Of all the nations considered in this study, only Nigeria offers a market of sufficient size and wealth to engender competitive struggles between a large number of firms. For most others, the present industrial order could be not a prelude to growth but a framework for economic stagnation.

PART II

Interpretation

CHAPTER 5

The Market as Political Arena and the Limits of Voluntarism

When African governments intervene in markets, they often do so in ways that harm the short-run interests of most farmers. On the one hand, by sheltering domestic industries from competition, they increase the prices that farmers must pay for goods from the urban areas. On the other, through the use of state power, they lower the prices that farmers receive for their products; alternatively, they compete with them in supplying food to the urban markets. And the benefits of the subsidies they do confer on farm inputs are reaped by the richer few.

This pattern of public policy raises many questions. Clearly the most important is: How do the governments get away with it? In countries that are overwhelmingly rural, as in Africa, how can governments sustain policies that so directly violate the immediate interests of the majority of their constituents?

One answer is that governments have the power to coerce. Chapters Six and Seven will examine their use of force against those who oppose these policies, and will stress their willingness to suppress those who attempt to organize against them. Nonetheless, it has long been recognized that although coercion is the ultimate basis of power, it is not a sufficient basis for governance. Domination by force alone is difficult to sustain. In this chapter we therefore examine additional factors that help to explain the durability of these policy choices.

PRIVATE CHOICE AND PUBLIC POLICY

One key reason why rural dwellers do not organize in opposition to government policies is that they fear government reprisals. Another is that they have a less costly alternative: they can use the market against the state, thereby evading some of the adverse consequences of government policies.

The capacity of farmers to use the market to safeguard their interests is documented in studies of the supply response of African producers. Maitha, for example, notes the capacity of small-scale coffee growers in Kenya to withdraw from ventures that have become economically unattractive. His data show that in the face of declining prices, producers devote fewer resources to coffee production: they harvest less intensively and place fewer acres under production.

To measure producer responses to prices, Maitha estimates the price elasticities of acreage and yield over the period 1946–1964. The price elasticity of acreage indicates the percent change in acreage associated with a percent change in price; and the price elasticity of yield indicates the percent change in yield associated with a percent change in price. Were farmers to cut their losses by reducing production in the face of declining prices, then, of course, we would expect to find positive estimates of these price elasticities. Maitha's data confirm this expectation. According to him, they show that the short-run price elasticity of yield was +.644 and that of acreage +.204. We should also expect production to be more responsive to price in the long run, rather than in the short run. In the long run, fewer factors are "fixed." It takes years for a coffee bush to bear fruit, for example; and we do not expect farmers to uproot coffee plants in the face of adverse prices. Short-run production changes will therefore be smaller than long-run shifts in production. Estimates of price responsiveness over the long run should therefore be greater. Maitha's data confirm this expectation as well.

Once again, he finds both elasticities to be positive, that of acreage being +.511 and that of yields approaching unity. Maitha's findings are replicated elsewhere in Africa (see Askari and Cummings).

African producers not only withdraw from ventures that have been rendered unattractive. They also alter their production mix to take advantage of shifting relative prices, thereby moving into the production of commodities for which the returns have become more favorable by comparison. In analyzing this behavior, researchers have estimated what are called "cross elasticities." When the price of a particular crop increases, then we should expect farmers to shift resources out of the production of other crops and into the production of the higher-priced commodity. Conversely, and more pertinent to this discussion, when the price of a particular crop declines, we should expect farmers to shift resources into the production of other commodities whose prices appear more attractive.

The cross elasticities tell us to what degree these expectations are confirmed. They measure the percentage change in production of one crop associated with the percentage change in the price of another. And if our expectations about the behavior of farmers are correct, the estimated cross elasticities should be negative. Scattered evidence suggests that they are. Adesimi, for example, found that tobacco production in Nigeria had a short-run cross elasticity with respect to the price of grain of −.96 and a long-run cross elasticity of −1.32. Bateman found a strong negative relationship between cocoa production and the price of coffee in Ghana over the period 1945–1963. Bateman's findings are corroborated by the work of many others (see Kotey et al., and Askari and Cummings). And in Zambia, changes in the price of tobacco and groundnuts compared to maize have led to a widespread decline in the production of these cash crops as farmers shift into maize production.

Along with the evidence of a positive "own-price" response, these data thus suggest that African peasants move out of the production of a crop whose price is on the wane and into the production of crops whose prices have become relatively attractive by comparison. By thus exploiting the alternatives open to them in the market, the peasants are able to defend their incomes against adverse shifts in the prices of particular commodities.

Rural dwellers have other alternatives. In particular, they can use the market for labor to defend themselves against the market for products. In the case of Kenya, for example, Huntington found that the decline in average earnings in the place of origin led to a large out-migration of persons; and an increase in average earnings in the place of destination led to a large and significant increase in in-migration. In both cases, the elasticity of labor supply exceeded 1. Barnum and Sabot (1977) replicated Huntington's study in Tanzania; and although they found lower elasticities, their analysis confirmed the ability of African populations to use the labor market to exit from areas where economic conditions have declined and to enter areas where the economic conditions are more favorable by comparison. Separate studies by Knight and Beals and Levy replicate these findings in Ghana. Of all the phenomena discussed in this book, rural-urban migration is the best researched. And the consensus is strong that economic incentives govern migratory behavior, and that African rural dwellers enter the urban labor market in search of increased incomes (see Brigg 1971, and Byerlee 1972; also bibliography in Bates 1976).

The capacity of African rural dwellers to exploit alternatives available to them in the marketplace is thus well established. What is particularly interesting is their ability to do so in efforts to avoid the depredations of the state.

One example comes from Tanzania. In 1974–1975, the government of Tanzania proclaimed the existence of a crisis in food production. The nature of this crisis is instructive. Food production in Tanzania may well have declined in the mid 1970s, in part because of drought and in part because under the government's so-called Ujamaa policy, farmers were compelled to abandon established farms and to move into villages. For whatever reason, it was the case that food prices rose; but the prices offered by the government marketing agencies did not. In the case of maize, the price offered by the government remained at 26 pence per kilogram between 1970 and 1973, rising to 33 pence only in 1974; this represented 50 percent of the world market price for the product. For wheat and rice, the prices remained virtually unchanged through the early 1970s and bore a similar relationship to the world market price (Tanzania 1977c; IBRD 1977). The records of the marketing agencies disclose a dramatic decline in food purchases: from 106 thousand tons of maize in 1971–1972 to 24 thousand in 1974–1975, and from 47 thousand tons each of wheat and rice in 1972–1973 to 15 thousand tons each in 1974–1975 (Tanzania 1977c).

The government interpreted this decline as a shortfall in production. Others, such as Hyden, interpreted it as a precapitalist reaction against the market on the part of peasant producers (Hyden 1980a, 1980b; see also Lofchie 1978). Reports from local field observations, however, fail to document significant declines in production, save those occasioned by local drought; and persons conducting fieldwork at the time recorded no such flight from the market. Rather, they recorded a flight from the government-controlled market, and a massive diversion of produce into private channels of trade (Frances Hill and William Jones, personal communications).

In addition to altering their market strategies, peasants also alter their production mix so as to avoid the burdens that governments impose upon them. This is particularly the case for the producers of cash crops, whose markets are more easily policed and controlled and whose prices are therefore more easily affected. In the Gezira scheme in the Sudan, for example, the government imposed a tax on the tenants to cover the costs of irrigation and technical services. This tax is collected from the proceeds realized from sales of cotton, and the tenants have responded by moving out of the production of cotton and into the production of untaxed commodities (DeWilde). A similar pattern was noted in the early 1970s in Senegal. Like many other governments in Africa, the government of Senegal controls the marketing of its principal export crop, which is groundnuts. In the late 1960s the government reduced the posted price of groundnuts by 15 percent; it also instituted new measures, such as a system of delayed payments, that reduced the actual price below the posted price. One response was illegal marketing: thousands of tons of groundnuts were smuggled annually through the borders with Gambia and Guinea. Another was a massive shift out of groundnut production and into the production of food crops. The shift in production characterized what came to be known as *le mal-*

aise paysan (see Schumacher; also Donal Cruise O'Brien 1979; and *West Africa*, March 10, 1980).

A final example is afforded by Ghana. For years Ghana has maintained a local price for cocoa that lies below the world market price (see Appendix B). The peasants have smuggled cocoa to Togo and the Ivory Coast, where more favorable prices are offered. They have also moved resources out of cocoa production. Thus reports document that within the cocoa-producing regions, labor is increasingly scarce; fewer workers remain in the industry, and those who do tend to be persons of greater age (DeWilde). Moreover, fewer investments are made in cocoa production, with the consequence that the industry is characterized by an aging stock of trees (ibid.). Lastly, land that was formerly in cocoa production is now being cleared and devoted to the production of other crops. This trend has been most vividly documented by the Ashanti Cocoa Project—a project designed to rehabilitate the cocoa industry by clearing the forests of old and diseased trees and replanting them with higher yielding hybrids. Project reports note that farmers readily agree to the clearing of old cocoa trees; they also note a rapid decline of interest when the farmers are asked to participate in the second stage of the project, which is the replanting of their farms with new seedlings. Moreover, the reports note that farmers seek to have their farms clear-cut; this represents a radical shift away from traditional forms of husbandry, in which shade trees were left to provide cover for the new cocoa trees. To the members of the Ashanti Cocoa Project, the motivation behind the farmers' behavior is clear. As their report for 1976–1977 declares: "[The] farmers desire to plant food crops rather than to plant and maintain high yielding cocoa. This, therefore, greatly influenced [the] method of land preparation. . . . It was observed that most farmers in the Project Area find food crop farming more attractive than cocoa farming and therefore do not take much, if any, interest in cocoa" (Ghana 1978, p. 33).

The marketplace offers several alternatives. The peasants can dispose of their crops through competing marketing channels. They can abandon production of a crop when its price declines and begin producing one whose price remains attractive by comparison. Or they can leave the industry entirely and enter other sectors of the economy. Given these alternatives, the peasants dodge and maneuver to avoid the deprivations inflicted upon them by public policy. They use the market against the state.

But this analysis should not be heralded as a triumph for the peasantry. The fact is that the peasants avoid the state by taking refuge in alternatives that are clearly second best. They move out of the production of the crops that are most profitable and into economic activities that have become more profitable only because they are less heavily taxed. In thus changing the way they employ their resources, they incur economic losses.

COLLECTIVE ACTION

The marketplace offers only private "solutions" to the collective problem confronting farmers, alternatives that can be chosen by individual farm families. As we shall note later, in Chapter Seven, farmers often do not dare to seek public solutions; for governments punish those who seek to organize in collective opposition to their policies. But even if political action were a more attractive alternative, most rural dwellers in Africa would operate at a relative disadvantage, because the structure of their industry presents greater obstacles to organizing than are found in other sectors.

Under most circumstances, producers want a higher price for their product. But a market price is not something that can be secured in the same manner as, say, a new bicycle. For if one person manages to get the price for a product set at a higher level, it is difficult to prevent other producers from also enjoying the advantages of that new price; they can either increase their profits by selling at the new price, or by undercutting it and thereby increasing the volume of their sales. As a consequence, efforts to alter prices are difficult to organize. Each producer has an incentive to free-ride: to let others expend resources in securing a change in price and then enjoy the benefits of the new price for free.

Nonetheless, it is obvious that groups successfully lobby for higher prices. It is also obvious that some groups organize more successfully than others. In Africa, as we have seen, prices are set in a way that favors the interests of industry but harms the interests of agriculture; and, within agriculture, prices tend to be more favorable for large farmers than for peasant producers. This suggests that certain factors operate to lower the costs of collective action for certain kinds of producers.

One major factor that influences the ability of groups to organize is the size distribution of their industry. The fewer the number of firms and the larger their individual output, then the smaller the incentives to engage in free-riding. When there are only a few major producers, the output of each at the higher price may more than cover their costs of lobbying to secure that price. And when there is only one producer of a product, there is of course no incentive to free-ride in efforts to raise prices. The costs of lobbying also vary with the size structure of the industry. When there are a few centrally located producers, the costs of communicating, negotiating, and coordinating strategies are comparatively low. But when producers are numerous and widely scattered, the costs of organizing are higher. The benefits as well as the costs thus vary, and the incentives to lobby are strongly influenced by the industrial structure.

This discussion is of obvious importance in explaining the relative success of industry as opposed to agriculture in securing policies that support their prices. As we have noted, the manufacturing sectors of many African nations contain less than a thousand firms; in any particular industry the number of firms can be small, often less than ten; and industries in many countries are simply monopolized by single firms. In each industry, very few firms tend to produce a large proportion of the output. The contrast with the agricultural sector is striking. In Ghana, Nyanteng estimates that there were 805,200 farm-holders in 1970 and 859,214 in 1974 (Nyanteng 1979, pp. 51–52). Zambia in 1969 had a rural population of 2,033,000, which suggests the existence of between 300,000 and 500,000 farm families, depending on the assumptions made about the average size of rural households (Bates 1976). In Kenya, the number of rural holders was estimated at about 1,500,000 in 1975 (Kenya 1977, p. 135). In Tanzania, there were approximately 2,500,000 farms in 1972 (IBRD 1977). Given the overwhelmingly rural nature of the African nations, the number of farm families is thus very large. And by contrast with urban industry, each rural producer generates a small proportion of the total output. The structure of the agricultural industries in Africa is suggested by the data in Tables 6 and 7.

Given the differences in the size distribution of rural as opposed to industrial producers, we should expect agriculture to stand at a relative disadvantage in organizing collective efforts to defend its interests. Evidence concerning the formation of interest groups in Africa is pertinent here. As part of a study of policy formation in the Ivory Coast, Michael Cohen (1974) analyzed President Houphouet-Boigny's initiation of public dialogues with major interests in that nation. In response to the President's initiative, various groups rapidly formed to articulate their needs in the policy-making process. They included midwives, nurses, health workers, gas station managers, school teachers, tailors, taxi drivers, tenants, construction workers, butchers, masons, trade union members, whole-

Table 6
Distribution of Size of Holdings and the Number of Holdings Relative to Cultivated Areas, Ghana

Size of holding (acres)	Percentage of farms
0–1.9	31
2.0–3.9	24
4.0–5.9	13
6.0–7.9	9
8.0–9.9	5
10.0–14.9	7
15.0–19.9	4
20.0–29.9	3
30.0–49.9	2
50.0 or more	2
Total	100

Source. USAID. *Development Assistance Programs for the Years 1976–1980, Ghana*, Volume 4, Annex D—Agricultural Sector. Mimeographed. January 1975, p. 76.

Table 7
Percentage Distribution of Holding Size by Rural Households, Kenya 1974–1975

Size	Percentage
Below 0.5 hectares	13.91
0.5–0.9 hectares	17.92
1.0–1.9 hectares	26.99
2.0–2.9 hectares	15.11
3.0–3.9 hectares	8.89
4.0–4.9 hectares	7.22
5.0–7.9 hectares	6.50
8.0 hectares and over	3.47

Source. Republic of Kenya. Central Bureau of Statistics. *Statistical Abstract*, 1977, p. 135.
Note. Figures are for holdings in areas excluded from former European highlands.

salers, and pharmacists. Cohen mentions the formation of a group of buyers of agricultural products, but not a group of producers, who apparently remained unorganized.

Further evidence comes from the list of the petitions submitted to the Adebo Commission in Nigeria. Of a total of 627 written petitions submitted to the Commission, only two came from organizations whose titles readily identified them as groups representing farm producers. Nineteen were submitted by organizations of agricultural workers, but none were submitted by persons who claimed to speak on behalf of independent farmers (Nigeria 1971, Appendixes 3 and 4).

Farmers thus seem to be relatively inactive in interest-group politics. The relative benefits and costs of collective action favor persons who work in other industries. And in the competitive struggle to manipulate the prices paid and received by farmers in the markets that determine their incomes, it is other interests who succeed in using the power of the state to alter prices to their advantage.

The Large Farmers

Within agriculture, however, not all farmers are treated equally. Large farmers, as we have seen, receive more favorable treatment than others; they are a powerful elite among rural dwellers. Their position illustrates the utility of the arguments that have been used to explain the general pattern. This section briefly examines two groups of large farmers: the Ghana rice farms and the farm lobby of Kenya.

Ghana Rice Farms. Whereas traditional rice farms average six acres in size, the commercial farms that arose in Ghana in the 1960s and 1970s tend to average over fifty acres or more. These large-scale farms are few in number; 400 to 500 is the figure most commonly given. Moreover, they tend to be concentrated in particular government "project areas." Because of their scale of production, they tend to dominate the market. As noted by Nyanteng: "The only crop [in Ghana] for which large scale production is of some significance is rice. In 1974, the large scale farm enterprises accounted for about one-third of the total acreage under rice production and a little over 50 percent of rice produced in that year. The trend in rice production shows that the dominance of small scale farming in the production of that crop is over" (1979, pp. 51–52).

On repeated occasions, the Ghanaian rice farmers have combined in efforts to drive prices up by withholding crops from the market. As Rothchild notes:

Rice farmers, disappointed over [their] incomes, protested against the 1974 prices of ₵ 14.70 per bag [14.70 *cedis* per bag]. When informed that they would be given an additional ₵ 1.00 bonus per bag . . . the farmers dragged their feet. Government statements on the extent of subsidies of fertilizer, use of combines at low cost, commercial loans, irrigation, the construction of feeder roads, and so forth did little to encourage the farmers to bring their rice to the mills. [Eventually] the Authorities . . . seized thousands of bags of paddy rice being "hoarded" by recalcitrant farmers. The rice farmers' protest against the fixed government price for their produce had been overridden, but not without the farmers having made their point to all observers of their country's affairs. [Rothchild, 1977, pp. 2–3]

Table 8
Changes in Size Distribution of
Large-Scale Farms, Kenya

Number of hectares	1954		1971	
	Number of farms	Percent of farms	Number of farms	Percent of farms
49 and under	467	14	741	23
50–399	1,162	37	1,253	40
400 and above	1,535	49	1,182	37
Total	3,164	100	3,175	100

Source. S. E. Migot-Adholla, "Rural Development Policy and Equality," in *Politics and Public Policy in Kenya and Tanzania*, ed. Joel D. Barkan with John J. Okumu. New York: Praeger Publishers, 1979, p. 161.

The organized withholding of the crop may have been overcome by the government's forces, but the effort of the rice farmers to influence prices has borne fruit. Underpinned by government subsidies for seeds, fertilizers, credit, and mechanical equipment, and by government support for ready access to "unused" lands, the rice industry has come to enjoy a pattern of protection which compares favorably with that conferred upon urban manufacturers in Ghana (Stryker).

The Kenya Farm Lobby. The efforts of the Kenyan government to engage in land reform are well known. The post-independence settlement schemes helped to move 500,000 people onto over 1,500,000 acres of land in the former white highlands; by so doing, they helped to assuage the land hunger that fueled so much of the nationalist movement. Less well known is the fact that 80 percent of the former white highlands were left intact and that the government took elaborate measures to preserve the integrity of the large-scale farms (see Migot-Adholla; Leys; Njonjo). The result, as illustrated in Table 8, has been the perpetuation of the large-scale farms in the former white highlands.

These farms are small in number and large in size. And although they market a declining portion of the crops produced in Kenya (see

Table 9), they still marketed one-half the value of all crops sold in 1974. The large-scale farmers of Kenya readily combine in defense of their interests. One of their most important collective efforts is the Kenya National Farmers' Union (KNFU). As noted in its annual report, "the KNFU's main aim is to ensure that farmers' claims and problems are taken into account by Government and other organizations closely connected with the industry" (KNFU 1971–1972, p. 3). The organization appears to be remarkably successful in this effort.

The KNFU is organized by commodity groups and by areas. It has committees for such commodities as beef, dairy products, cereals, and coffee; and it also has committees for geographic areas. It organizes both large and small farmers. The small farmers, being less specialized in production, tend to participate mainly in the area committees, whereas the more specialized large-scale producers tend to work through the commodity committees.

It is clear that the KNFU is dominated by the large-scale farm-

Table 9
Share of Gross Marketed Agricultural Production Between Large and Small Farms, 1958–1974, Kẹnya

| Year | Large Farms | | | Small Farms | | Total |
	Value (K£M)	Percent	Value (K£M)	Percent	Value (K£M)
1958	33.4	81.0	7.6	18.0	41.02
1959	33.9	80.0	8.4	20.0	42.30
1963	40.9	78.0	11.6	22.0	52.50
1968	34.4	49.0	35.8	51.0	70.20
1972	50.3	47.5	55.6	52.5	105.90
1974	73.4	49.4	75.0	50.6	148.40

Source. Apollo Njonjo. "The Africanization of the 'White Highlands': A Study in Agrarian Class Struggles in Kenya, 1950–1974." Ph.D. Dissertation, Princeton University, December 1977, p. 149.

ers. They fill its national offices and staff the commodity committees as well. And although they comprise less than 1 percent of the agricultural producers of Kenya, they account for more than 50 percent of the membership of the KNFU and pay in excess of 80 percent of its subscriptions (see KNFU *1971–1972*, pp. 1–7). One consequence is that the KNFU lobbies for programs that chiefly benefit big agriculture, which in turn creates a distinct bias in the allocation of public services in favor of large-scale farmers. This result has been documented by Leonard for the Kenyan extension services, and also by the government's credit program, which extends its benefits only to farms with 15 acres or more planted with cereal crops.

Nevertheless, it can still be argued that the KNFU helps to create a framework of public policies that provides an economic environment highly favorable to all farmers, whether small or large. Perhaps the most striking evidence for such a contention is offered by the setting of prices in 1976. In January of 1975, the government of Kenya announced a comprehensive price review for most major commodities. The KNFU submitted "detailed proposals and costings" (KNFU *1974–1975*, p. 9) but was disappointed by the government's action. The price of two commodities, wheat and maize, was not changed, and the price increases announced for other commodities were considered too little and too late to be of much benefit to farmers (ibid.).

The next year, the Union determined to do better. In September 1976, the KNFU "initiated a delegation to see His Excellency Mzee Jomo Kenyatta" and "a number of carefully prepared memoranda and proposals were prepared and submitted" (KNFU *1975–1976*, p. 9). The results "far exceeded" expectations (ibid.). Maize prices increased 23 percent; wheat, 20 percent; and beef between 15 and 23 percent, depending on the grade. And because consumer prices were not increased as well, many of the government boards that purchased and marketed these commodities incurred substantial losses and had to rely on government credits to remain solvent (see *The Standard*, October 20, 1979; *Daily Nation*, October 2, 1976). It is true that the benefits of this price increase go mainly to the large farmers, and that marketing arrangements would augment this effect. Nonetheless, the rise in prices was given to all producers, and the action of the large farm lobbyists therefore generated benefits for the industry as a whole.

The cases of the rice industry in Ghana and the KNFU in Kenya thus illustrate the ability of large-scale farmers to act collectively in defense of their interests. Being large in size, each stands to reap substantial benefits from higher prices. And being small in number and often geographically concentrated, the large farmers face relatively low costs of organizing. Small farmers, by contrast, are numerous, and widely scattered, and each of them markets a small volume of output. Efforts to organize in support of higher prices thus tend to be more costly in the small farm community and to offer fewer private advantages. Insofar as public policy reflects the lobbying efforts of the citizenry, it therefore tends to favor the large farmer.

The cases illustrate another point. The interests of the large and the small farmer often conflict. As argued in Chapter Three, the gains secured by the one are often at the expense of the other. Nonetheless, price rises for crops benefit all farmers, though not to the same degree. The actions of the large farmers to produce public policies in support of higher prices result in a more favorable structure of prices for agriculture as a whole.

There are two important implications. The first is that as the number of large farmers increases in Africa, the farming community will tend to grow politically more assertive. Conflicts between producer interests and the interests of others will intensify as the large producers give political weight to the economic demands of farmers. The second is that countries with greater numbers of large farmers will tend to have agricultural policies that offer more favorable prices to farmers. The Ivory Coast and Kenya are cases in point. Planters, large farmers, and agribusiness in the two countries have secured public policies that are highly favorable by comparison with those in other nations. Elsewhere the agrarian sector is better blessed by the relative absence of inequality. But it is also deprived of the collective benefits which inequality, ironically, can bring.

CHAPTER 6

Rental Havens and
Protective Shelters:

*Organizing Support Among
the Urban Beneficiaries*

The new nations of Africa were born in a moment of hope. It is difficult to recapture the emotional tone of that moment. But the depth of it, the fullness of it, and the promise it offered left its mark on all who were in any way touched by events of that era. It was called a new dawn, a rebirth, a reawakening.

For many, the dreams of that period have given way to disillusion. Social scientists studying the United States long ago learned to listen to the "little man from Missouri." The sullen cynicism of the common man of Africa today offers no less insight into the reality behind the public-spirited rhetoric of the policy process. Public institutions no longer embody a collective vision, but instead reinforce a pattern of private advantage that may often be socially harmful—that is the message of disillusion in Africa today.

In re-evaluating the promise of the nationalist period, many have charged its spokesmen with cynical manipulations of popular hopes and with self-interested proposals of programs of questionable merit. There is much to this interpretation, but it captures only part of the truth. For during the nationalist period, there were in fact striking instances of public spiritedness. People made major sacrifices for the sake of national independence. Careers were abandoned, educations sacrificed, and lives lost as persons turned from normal pursuits and entered the political arena. And many did so for high-minded reasons: to get rid of foreign rule, to end racial oppression, and to escape colonial bondage. Above all, they did so in order to seize control of the state. They sought to secure for the people of Africa the power to create and implement public policies and thereby secure greater prosperity.

The words of Kwame Nkrumah—"seek ye first the political kingdom and all else shall be added unto it"—best represent the vision of that era. And in keeping with Nkrumah's injunction, public servants sought to use the power of the state to manipulate major markets and thereby induce a flow of resources that would generate rapid development.

The forms of economic manipulation chosen were compatible with prevailing economic doctrines. Many of those who formulated and implemented the development programs of the new African states had studied the theories of the leading development economists. That industry is the engine of growth; that savings come from the profits of industry and not from the profits of farmers; that resources should be levied from the countryside and channeled into industrial development; that the rural sector should be squeezed for development and can be made to give up resources without major declines in production—these were and remain today important tenets in development doctrine (see Lewis; Ranis and Fei; Jorgenson; and readings in Stiglitz and Uzawa).

It would therefore be a mistake to see the policies chosen by governments in Africa as representing commitments made without regard for the public interest. But what is notable is that the mix of policies chosen to secure economic development has permitted the entrenchment of enormously powerful private interests, and that this fact has become an important source of the durability of policy commitments.

PUBLIC POLICY AND PRIVATE ADVANTAGE

The dynamics are simple but powerful. The government enters certain markets. For development purposes, it lowers prices in those markets. With lower prices, demand increases; private sources of supply furnish smaller quantities; and scarcities therefore occur under circumstances of excess demand. The result is that the commodity in question—be it foreign exchange, capital for investments, or whatever—achieves new value. Insofar as a public institution controls the market for that commodity, it then has control over this new value.

The administrators of such an institution can consume that value themselves, which is financial corruption. Or they can apportion it to others whose influence they wish to secure, which is political corruption. In either case, the bureaucracy that is mandated to control the operation of a market for public purposes finds itself in control of financial and political resources—resources that render the program economically useful to those in control of it and a means for generating a political following. Market intervention leads to the formation of vested interests in policy programs.

The process is outlined in Figure 2. P_0 represents a price at which the market is in equilibrium; at P_0 the quantity demanded (Q_{D_0}) equals the quantity supplied (Q_{S_0}). Assume that the government, desiring some policy objective, intervenes in the market and lowers the price to P_1. At this lower price, consumers demand more of the good, so the quantity demanded increases $(Q_{D_1} > Q_{D_0})$. But at the lower price producers will supply less of it, so the quantity supplied declines $(Q_{S_1} < Q_{S_0})$. At P_1 the quantity demanded therefore exceeds the quantity supplied $(Q_{D_1} > Q_{S_1})$ and the market cannot allocate the good; too little is available for the level of demand at the new price. Rather, the good will have to be given to some and withheld from others who want it at that price. It will have to be rationed.

Being subject to excess demand, the good increases in value. At the quantity supplied (Q_{S_1}) it is scarce by comparison with the demand for it. As seen in Figure 2, some consumers would be willing to pay P^* for the good. P^* lies above P_1, the officially mandated price; it also lies above the market clearing price, P_0. The difference between P^* and P_0 can be regarded as a premium created by the scarcities induced by government intervention. We will call this

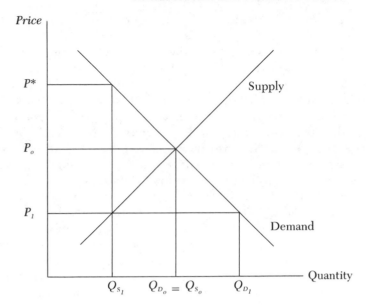

Figure 2.
Excess Demand in a Market

premium an *administratively generated rent*: a value in excess of the market value which has been created by an administratively generated fixity in the supply of a commodity (see Krueger; Posner).

The value of this rent can, of course, be appropriated in the form of bribes; or those in charge of the market can confer it upon others by giving them rations of the commodity at the administratively lowered.price. In the latter case, the bureaucracy creates grateful clients—people who owe their special fortunes to public officials who choose them, from among competing claimants, for privileged access to these resources. Government intervention, excess demand, and the conferral of privileges are thus all part of the political process by which public programs create vested interests in policies of social and economic reform.

The Urban Sector

These dynamics characterize the operation of programs designed to promote the development of urban-based industry by extracting capital from export agriculture. This is best illustrated by the material from Western Nigeria. There, as we have seen, the marketing board lowered the prices offered peasant producers for export crops and thereby accumulated surplus revenues. A portion of the proceeds thus generated by the board was transferred to development agencies, which provided capital for loans at subsidized terms to potential investors in the urban industrial sector.

As our analysis would suggest, this policy in Nigeria was in fact self-contradictory, and its contradictions became a source of political opportunity. One of the results of making capital available at lower prices was to create a "shortage" of capital; artificially lowering the price of capital led to excess demand for it (Schatz 1977, pp. 66ff). And it was precisely at the time of the capital shortage that characteristic loans were made—loans whose beneficiaries were members of the agencies themselves, or politically influential persons whose support the agency heads wished to secure (Schatz 1970, pp. 41ff).

Further evidence (much of which is cited by Schatz) is contained in the volumes of the Coker Commission of Inquiry into the affairs of statutory corporations in Western Nigeria. There we find that the persons in charge of the development agencies used their powers to secure the transfer of funds into banks and corporations in which they themselves held directorships. In effect, they consumed the administratively created rent themselves. As directors of the banks they gave themselves large, unsecured, interest-free loans. In the words of the Commission, one witness "told us of how between the years 1958–1959 he received a total amount of over £1 million from the National Bank without signing any papers for the amounts" (Nigeria 1962, vol. 2, p. 7). And as directors of the corporations, these persons secured both fees and profits.

The ability to ration capital not only led to personal gain, but was also employed to build political coalitions. In evidence of this, one of the major Nigerian corporations that secured subsidized loans, the National Investment and Properties Company, published the newspaper chain owned by the Action Group, the party that held power in the Western Region. The Commission of Inquiry noted that the company and the party were virtually identical (Nigeria 1962, vol. 1, p. 55). Furthermore, major corporations owned by politically influential persons, including the "father" of the Action Group, received major loans during the period of supposed capital scarcity.

In their efforts to promote the growth of urban-based industries, governments in Africa not only intervene in the markets for capital; they also artificially increase the value of the domestic currency, thereby cheapening the costs of capital equipment which is scarce domestically and must be imported from abroad. The result of the maintenance of an overvalued currency is, once again, the creation of an excess demand for foreign goods and the elaboration of means for rationing access to them. Financial corruption and the apportionment of privileged access are once again correlative results; and both create private incentives for persons to support the continuation of this policy measure.

Perhaps the best known examples come from Ghana. Following the "big push" of the post-independence industrialization programs, Ghana began to run large deficits in foreign trade; its massive international reserves which had been accumulated from cocoa exports began to erode, and "by November 1961 the government felt it had only one instrument left to deal with the situation: stringent import licensing" (Leith, p. 23). The evidence strongly suggests that those in charge of the regulation of foreign exchange rapidly converted the scarcities in this market into personal wealth. The first minister in charge of the program, Mr. A. K. Djin, owned a trading firm which, by privileged access to import licenses, grew from a minor corporation into one of the major import houses in Ghana. His successor, Mr. Kwesi Armah, also secured major personal benefits. As noted in a government report: "He introduced the system whereby all applications for import licenses had to be addressed to him personally under registered cover and he alone was responsible for processing the said applications. . . . Import licenses were issued on the basis of a commission corruptly demanded and payable by importers on the face value of the import licenses issued. The commission was fixed at 10 percent, but was in special cases reduced to 7.5 or 5 percent" (Ghana 1966, p. 26).

As suggested in this quotation, the value of access to quantities of foreign exchange was not consumed solely by those who administered it; it was also apportioned to others. Kwesi Armah, for example, gave the right to negotiate with petitioners for licenses to selected Members of Parliament; they then became his protégés, with grateful clients of their own (Republic of Ghana, 1967, pp. 4–5). And after the overthrow of the Busia government, the military government found that the Minister of Trade "had varied the normal procedures for allocating licenses to favor specific individuals and companies who were [backing] the ruling party" (Killick, p. 281).

Thus far we have indicated how government intervention, by depressing prices in markets, creates opportunities for conferring privileged access to commodities that have been rendered scarce in comparison to the demand for them. Privileged access is used by the elites in charge of the programs for direct personal gain or to create a political following. The political attractions are obvious. And they help to explain why, when given a choice between market and nonmarket means for achieving the same end, African governments often choose interventionist measures.

Leith, for example, notes that the foreign exchange crisis in Ghana in the 1960s could have been resolved either by devaluation or by import licensing. Ghanaians were demanding more foreign exchange than was being supplied by exports; their currency was overvalued and there was thus an excess demand for foreign imports. As Leith points out, the government could have devalued; by raising the price of foreign currency, it could lower the quantity demanded. Alternatively, the government could employ a system of rationing. Leith then states: "The immediate . . . difference between the two [approaches] was nil. Curiously, though, a given volume of foreign-exchange use at a lower *cedi* price to the initial recipients seemed preferable to the same volume at a higher *cedi* price" (Leith p. 156).

In our analysis, an appreciation of the *noneconomic* difference between the two approaches dissipates any puzzlement over the preference for administrative controls. Those who received the lower-priced foreign exchange were given special favors, and those who apportioned it amassed a political following. To have allocated foreign exchange through the market would have given no comparable chance for the exercise of discretion, and thus no comparable opportunity for creating a political clientele.

URBAN INDUSTRY, PUBLIC POLICY, AND POLITICAL POWER

In earlier chapters we have seen how governments create highly sheltered markets for urban-based industries. Protected by government policies, these markets confer benefits upon those who produce within them. The benefits take the form of noncompetitive prices, which generate *noncompetitive rents*: they are increases in the earnings of firms created by the ability of prices in the protected industry to rise above the level that would be sustained if the industry were subject to competition. Like the administratively generated rents, these rents too are consumed by the elite or distributed to the politically faithful.

It will be remembered, for example, that in order to secure Firestone's investments in a tire factory, the government of Kenya gave Firestone a virtual monopoly over the tire market for ten years, sheltering it from both domestic competition and from foreign imports. This agreement virtually guaranteed Firestone monopoly profits. It also generated advantages to public officials and their political allies. As part of its agreement, the company shared with the government the right to name its distributors; the government could then pick those persons who could share in the monopoly profits. The company also brought prominent Africans, including one former Cabinet Minister and "one of the chief negotiators" on the Kenyan side of the deal, into managerial positions in the sheltered firm (Langdon, p. 173).

As a World Bank report states, the system of protection in Kenya "created absolute protection for many manufacturers of consumer goods. On the basis of the mission's interviews *it would not be an exaggeration to suggest that several firms have a license to print money*, being subject to no competition at home or abroad" (IBRD 1975, p. 298, emphasis added). These profits are shared with the administrators of the public policies that helped create them. Cohen and Swainson document the tendency of the political elites in the Ivory Coast and Kenya, respectively, to hold directorates in private firms and state industries; Shivji does the same for permanent secretaries in Tanzania (p. 89). Moreover, the policy-generated rents are also apportioned among political allies. Arthur Lewis, for example, in reviewing the state industries of Ghana, wrote that they have "suffered greatly from outside interference, in the shape of members of Parliament and other influential persons expecting staff appointments to be made irrespective of merit, redundant staff to be kept on the payroll, disciplinary measures to be relaxed in favor of constituents" (quoted in Killick, p. 245). When attacked for using the state industries to provide sinecures for political allies, N. A. Welbeck, a minister and sometime Secretary General of the ruling party in Ghana, simply replied: "But that is proper; and the honorable Member too would do it if he were there" (quoted in Killick, p. 245). That subsequent regimes have behaved in the same way only underscores the sagacity of Welbeck's reply.

Inflated payrolls are a characteristic of many state industries in Africa, as elsewhere. In part, this is another indication of the artificially inflated prosperity of the urban industries. But it is also indicative of the political uses of these policy-induced rents. Protective measures inflate the profits of firms. These revenues can be shared with political clients. And recruitment to jobs in state industries becomes a basis for allocating these profits so as to form political organizations.

In his study of the membership of the boards of Ghanaian state enterprises, Henry Bretton, 'for example, noted that they "had been staffed not with the most competent but with the . . . friends and associates of the President," that "thousands of employees . . . were at the President's mercy and disposal," and that "the efficiency-oriented staff members in the Secretariat who attempted to halt the drift to institutionalized incompetence and corruption were fighting a losing battle" (quoted in LeVine, pp. 74–75). Less dramatically, perhaps, but in accordance with the same logic, is the conduct of Daniel Arap Moi, President of Kenya. In an effort to consolidate his power after succeeding Jomo Kenyatta, Moi made series of high-level appointments to state corporations. These appointments were overtly political and were made in the wake of the Ndegwa Commission's condemnation of the inefficiency of these bodies, and its call for their reform (*Weekly Review*, May 16, 1980, and May 23, 1980). That Moi would choose to do this emphasizes the overriding importance of political considerations and suggests the ways in which economic inefficiency can be used for political purposes.

A more egregious example is provided by the rule of President Mobutu in Zaire. An important basis of Mobutu's power is his capacity to make appointments to state-regulated industries. Efforts at economic reform repeatedly founder on his determination to manipulate these industries in order to generate privileges for himself and his followers, and to reward those whose support he needs to remain in power. And in Liberia, when Master Sergeant Doe began his "revolution" the heads of government firms were prominent among those listed as exploitative members of the old order (*Los Angeles Times*, April 26, 1980). A key signal of the limited nature of Doe's rebellion in the eyes of many was that the lives of these persons were spared. The revolution did not, as it had promised, break up the pattern of privilege characteristic of the "old Liberia"—a structure of advantage and power based in large part on the consumption and distribution of rents in state-dominated markets.

By intervening in markets for capital and foreign exchange, and by influencing the structure of markets for manufactured items, the governments of Africa have sought to use government power to promote urban industrial development. Industrial development is equated with the public good. These policies create economic environments which generate rents. The rents are both economically valuable and politically useful, and from them are forged bonds of self-interest that tie African governments to their miniscule industrial base. Thus policy choices, made to serve a new vision of the public good, have created a network of self-interest which has proved more enduring than the faith which that vision initially inspired.

CHAPTER 7

The Origins of
Political Marginalism:

Evoking Compliance
From the Countryside

In Africa, as elsewhere, governments use force to quash peasant resistance to measures intended to create a new political and economic order. By frustrating those who would seek fundamental changes, governments remove proposals for comprehensive reforms from the political agenda and forbid organized efforts to alter the collective fate of the disadvantaged. Instead, they allow only efforts to seek marginal adjustments to the status quo, or petitions for individual exceptions to it. The capacity to coerce is thus used to defend and perpetuate basic policy commitments and the political and economic order they create.

Among the primary objects of government coercion in Africa are opposition parties. In this book we are concerned with parties that mobilize rural populations against the agricultural policies of governments in power. African governments use their control of the courts and the legal system to harass such parties, to ban them, and to arrest and imprison their leaders.

We have already noted the effort of the Government of Ghana to counter the appeals of the National Liberation Movement (NLM), which was organized in the cocoa belt to oppose the government's cocoa pricing policy. Accounts of political life in rural Ghana during the struggle with the NLM are filled with accounts of roadblocks, beatings, assassinations, and clashes between armed groups of party militants (Dunn and Robertson; Owusu). What was decisive in the end, however, was the government's control over the police and the courts. The opposition party was banned, its activities proclaimed illegal, and its organizers arrested for "reasons of state."

A similar tale emanates from Kenya, where in the 1960s a dissident faction of the ruling party, the Kenyan African National Union (KANU), opposed the government's program for establishing private rights in land. This faction correctly claimed that the program tended to confer disproportionate benefits on the wealthy, who could afford to buy land, while failing to safeguard the assets of the poor, and in particular the former freedom fighters. Because demands for land had furnished much of the impetus for the struggle for independence in Kenya, this critique was politically telling. To forestall damage to their political standing among the militant rank and file of the governing party, members of the dominant faction therefore altered the party's constitution. The changes they made enabled them to remove from party office the principal spokesman for the dissidents. The dissident faction thereupon split from the KANU and joined an opposition party, the Kenya People's Union (KPU). By controlling the police and the courts, KANU was able to frustrate the growth of this opposition party. In the local government elections of 1968, for example, the judiciary found technical faults in all but six of the nomination papers of KPU candidates; none of the papers filed by candidates for KANU were found defective (Leys, p. 216; Buijtenhuijs). Then, after clashes between KPU and KANU supporters, including one in which the President's bodyguard fired into a crowd of KPU loyalists, the KPU was banned and its leaders detained. Suppression of the KPU put an end to attempts to change the government's agricultural policies by organizing a political movement capable of removing incumbent elites from office.

The fate of the NLM in Ghana and the KPU in Kenya is paralleled by the fate of opposition parties in a host of other African nations. Because the majority of the African people live in rural areas, it is inevitable that their fate becomes central in the appeals of any political opposition. With the use of the state's instruments of coercion to emasculate the political opposition, governments in power thus eliminate one of the basic elements of political life which, by the sheer weight of self-interested political calculation, would champion the interests of the rural majority.

Instruments of coercion are also used against political entrepreneurs who seek to build their careers on protests against agricultural policies. An example would be J. M. Kariuki of Kenya. A former freedom fighter, Kariuki attained prominence in post-independence Kenya as a private secretary to President Kenyatta, a member of parliament, and a member of the government as well. Increasingly, however, Kariuki dissented from the government's position. The principal source of his disaffection was the tendency of members of the political elite to use government programs to acquire large agricultural holdings. Foremost among these new landholders, of course, were the President and his family. As described by one political observer, Kariuki "was not only pointing to the vast lands that every peasant believes, correctly, the Royal Family has acquired. He was inviting his audience to remember that Mau Mau had sprung from the land issue. Finally, he even said it plain: 'Unless something is done now, the land question will be answered by bloodshed'" (*Sunday Times*, London, August 10, 1975, p. 3).

The question did indeed lead to bloodshed, but the blood was Kariuki's own. In February of 1975, President Kenyatta's fields in Rongai were burned. His cattle were hamstrung. Leaflets were circulated describing the wealth of his family. Kariuki was suspected in having a role in organizing these activities. On March 1, he was abducted, taken into the hills outside Nairobi, and murdered. A subsequent Parliamentary investigating commission implicated persons close to the President: his bodyguard, his brother-in-law, his Minister of State, and his closest friend. For their efforts to uncover the truth about the murder of Kariuki, members of the commission were themselves jailed by the government of Kenya.

RURAL DEMOBILIZATION

Besides using their power to forbid collective efforts at altering the social standing of the peasantry, governments in Africa also use their control of markets to fragment the rural opposition. They accomplish this by making it in the private interest of individuals to cooperate in programs that are harmful to the interests of producers as a whole.

As we have seen, under many governments, all producers in the countryside are subject to a depressed price for their products; this is particularly the case for export crops. Through the pricing policies of government agencies, the public sector accumulates revenues. What is critical is that the governments return a portion of these revenues in the form of divisible benefits, which they confer upon supporters and withhold from political dissidents. The apportionment of these divisible benefits becomes a basis for attracting allies and building political organizations.

In part, this tactic underlies one of the most paradoxical features of agricultural programs in Africa: the coexistence of taxes (through the imposition of low prices on products) and subsidies (through the reduction of prices of factors of production). Deference to pricing policy is obtained by the manipulation of subsidy programs. The use of subsidy programs to build political support in the countryside for governments in power, and particularly for their agricultural programs, is a prominent feature of agrarian politics in Africa.

In Ghana, for example, in 1954 the Convention People's Party (CPP) government of Kwame Nkrumah passed the notorious Cocoa Duty and Development Funds (Amendment) Bill; it thereby froze the producer price for cocoa for four years, anticipating increased government revenues from the trading profits in this commodity. An immediate result was the formation, in the cocoa-growing regions, of a powerful opposition party, the National Liberation Movement (NLM), which opposed the new ordinance. The NLM threatened to unseat the CPP throughout the cocoa-growing regions, and so the CPP fought back. One of the resources at its command was the allocation of subsidized inputs. The government loan program, for example, became a weapon in the struggle to build a pro-government organization in the cocoa-growing region and to counter growing resistance to the government's pricing policy.

The "farmers wing" of the governing party was the United Ghana Farmers' Council (UGFC); keeping this in mind, the following comments in a government report are instructive. "The declared policy of CPC [the Cocoa Purchasing Company] not to grant loans to any farmer who is not a member of the UGFC is to be found at page 80 of the CPC Minutes Book (Exhibit 157), where it is laid down that before a farmer is considered for a loan, the officer dealing with the application must satisfy himself that the applicant is a bona fide cocoa farmer and that he is a member of the UGFC" (Ghana 1967a, p. 1). The report also noted that "the distribution of gammalin [an insecticide], cutlasses, etc., gave the United Ghana Farmers' Cooperative Council officials an opportunity to accord preferential treatment to their favorites and party members" (ibid., p. 18). Through the manipulation of these subsidized inputs, the government was able to erode support for the NLM and its opposition to the freeze in cocoa prices.

The use of government agricultural programs to organize rural support is also revealed in materials from Senegal. The government of Senegal secures much of its revenues from the marketing of groundnuts; and at least one quarter of the country's groundnuts are grown in areas controlled by an Islamic sect commonly known as the Mourides. The leaders of the Mourides are known as Marabouts. With the postwar expansion of the franchise in French West Africa to include residents of the countryside, the Marabouts gained in power. As Donal Cruise O'Brien (1971) states:

The result of postwar rural enfranchisement was that the *marabouts* became political agents for the major parties, whose feeble organizations were inadequate to reach the mass of ordinary peasants, and which were often unable to capture their interest in political programs which had little obvious relevance to the immediate problems of rural life. The easy way to win rural votes was through notables who could guarantee the votes of their followers, and of all the notables the most important were the *marabouts*. [P. 262]

The problem of securing the backing of the Mourides became more urgent with the advent of self-government in Senegal, for the government relied heavily on the profits of the groundnut trade to provide revenues for its expanded development programs. Maintaining a producer price for groundnuts well below the price realized from sales in the French market, the government of Senegal curried favor with the Marabouts by giving them privileged access to publicly subsidized inputs: fertilizers, mechanical equipment, land carved out from forest reserves, and above all massive amounts of government credit. The government, in short, has used its control over the allocation of subsidized farm inputs to build a political organization. Donal Cruise O'Brien, citing Brochier, refers to the end result as the creation of a "technically oriented feudality." As he states:

J. Brochier correctly points out that the projects involving technical co-operation with the *marabouts* have always reinforced the hierarchical structure of the brotherhood. The Government, by providing land, credit and various forms of technical assistance for the Mouride leaders, has in general contributed to the *marabouts'* means of domination, and has more particularly helped to bring about a concentration of resources in the hands of a few notables with significant political influence. [Cruise O'Brien 1971, pp. 227–228]

By conferring privileged access to subsidized inputs upon the Marabouts, the government thus enhanced their power over the rural masses; and by so doing, it helped to build its own political machine in the countryside of Senegal.

The cases just cited are to some extent exceptional. Nonetheless, the tendency to use control over farm implements to build organized rural support for governments in power is a pervasive one. In Zambia, the cooperative movement formed an important basis for the rural political organization of the United National Independence Party (UNIP), the governing party. It was through the movement that small-scale farmers could gain access to loans, seeds, fertilizers, and mechanical implements. And the UNIP's domination of the movement enabled it to apportion these benefits to the political faithful. Rationing access to farm inputs became a means of consolidating political power in the countryside (Bates 1976).

Moreover, the government credit agency in Zambia was heavily staffed by persons transferred over from UNIP. One former UNIP Regional Secretary whom I knew moved from the party to the credit organization. He regarded his work with the agency as a continuation of his career as a political organizer. And though he was professional enough not to want to "waste the government's money" by lavishing funds on political figures who were poor economic risks, he nonetheless regarded the money he could distribute as a useful tool for convincing the people of the beneficence and power of the governing party. The political use of loan funds has also been recorded in Nigeria (Northern Nigeria 1967, vol. 3, p. 6). And the capacity of the government of Tanzania to mobilize its population into Ujamaa villages has been attributed partly to its ability to link access to subsidized farm inputs to village membership (Raikes; McHenry).[1]

Through the use of violence, the governments of Africa have forestalled the mobilization of the rural majority against policies that harm their interests. And by granting or withholding farm inputs, they gain the backing of individual farmers for programs which, taken as a whole, do basic violence to the interests of agricultural producers. The agricultural programs of African governments thus become basic units of rural political organization.

But this organization is narrowly based—its members are the better-off few, and most of the peasantry remain outside it—and therefore vulnerable. During the nationalist period in central Africa, for example, those mobilizing the rural masses against the colonial regimes attacked the elite farmers who benefited from access to inputs whose costs were subsidized by the colonial governments. In many locations, the progress of the rural insurrection was measured in terms of the rate of defection of these elite farmers from government agricultural programs. (Interviews in Zambia, 1971–1972; see also Baylies; Dixon-Fyle). The progressive farmers, employing new technologies disseminated by their governments, are thus a frail base on which to build a rural constituency.

These organizations are also vulnerable because they are expensive. When governments are poor, when other programs win out in the competition for scarce resources, and when international organizations fail to contribute significantly to the costs of agricultural

1. It is notable that in the rural portion of their program in Afghanistan, the Soviet-backed government in Kabul offered the peasants "new farming tools, fertilizers, and liberal loans in an apparent effort to blunt rebel recruitment" (*Los Angeles Times*, February 29, 1980).

development, then governments must lower the level of the subsidies they provide farmers. At this point, the elite farmers lack the incentive to act as persons whose interests lie apart from those of the larger rural population. They may then offer leadership in organizing the collective opposition to the policies of the governments in power. The rural strategies of African governments are thus vulnerable to the vagaries of their fiscal base and the means by which they attempt to ensnare the larger farmers.

PUBLIC SERVICES AND PUBLIC PROJECTS: CONSTRUCTING A SYSTEM OF SPOILS

In their efforts to organize political support in the countryside, African governments also manipulate the structure and performance of their public services. Governments everywhere supply roads, clinics, schools, water supplies, and the like. In Africa, and in other developing nations, "development" projects are also standard fare. And whether it be in Mayor Daley's Chicago or Awolowo's Western Region of Nigeria, the supply of such services can be, and is, tailored to the quest for political support.

Studies of the behavior of members of parliament in Africa uniformly stress the emphasis they give to securing schemes and projects for their districts (Cliffe for Tanzania; Barkan for Kenya; Dunn for Ghana; Bates for Zambia). Moreover, studies of the attitudes of African electorates indicate that citizens seek, and expect to get, material improvements from those with access to public power (Barkan, Bates, and others). Holders of public office fully realize that in order to remain in power, they must manipulate the bureaucracy of the state to secure such benefits. The result is a general tendency to try to orchestrate public programs to secure political advantage.[2] And, of greater relevance to this work, the tendency is particularly strong with respect to agricultural programs.

2. Illustrative of the general tendency is Michael Cohen's analysis of the behavior of the Parti Democratique du Côte d'Ivoire (PDCI), the ruling party in the Ivory Coast. As he states: "The failure to express sufficient militancy for the PDCI and the government leads to neglect. . . . On the other hand, support for the government or one of its key figures is rewarded with the granting of public resources. Thus in 1962,

Several features of the agricultural programs of the states of Africa can be attributed to the quest for political support. One of these is the preference for production schemes as opposed to pricing policies in the attempt to secure greater food supplies. Another is the structure of these production programs—their number, their location, and their staffing.

Positive pricing policies are politically unattractive to African governments seeking greater food production. Their political costs are high in terms of loss of support in the urban areas; and their political benefits are low in terms of their ability to secure support from the countryside—or at least they are low by comparison with those which can be secured from the allocation of production projects. Were the governments of Africa to confer a price rise on all rural producers, the political benefits would be low; for both supporters and dissidents would secure the benefits of such a measure, with the result that it would generate no incentives to support the government in power. The conferral of benefits in the form of public works projects, such as state farms, on the other hand, has the political advantage of allowing the benefits to be selectively apportioned. The schemes can be given to supporters and withheld from opponents. Project-based policies, as opposed to pricing-based policies, are thus relatively attractive from the point of view of organizing a rural constituency in support of the government in power.

Governments can choose where to locate such schemes. They can also choose with whom to staff them. Both decisions offer opportunities for organizing political support.

The importance of political motivations is suggested in features of the state farm programs in Western Nigeria and Ghana. In both cases (reported by Wells and Dadson, respectively), the programs "over-expanded": state farms were provided for every electoral district! By most accounts, this decision crippled the programs from an

---

when [the] Minister of Construction and Town-Planning . . . had to choose one hundred villages to receive . . . improvements, he received a list of localities from [the] president of the National Assembly and PDCI secretary-general. The list had been drawn up by députés from all over the country in an effort to reward loyal populations and encourage support from opposition groups" (p. 90).

economic point of view; having so many farms meant that too few resources were provided for each, with the result that most operated inefficiently. But, from a political point of view, structuring the programs so as to provide a state farm in each constituency made available to government backers in each district public resources with which to organize support of the government in power. Moreover, within each district, the state farms were often poorly located, again from the point of view of maximizing production. A principal reason for this, apparently, was a desire to put them in areas where they would provide a "public works" benefit to the supporters of the government in power. As Wells states, in a bemused comment on Nigerian farming schemes: "allocations were used in an attempt to solve essentially political problems, often at the cost of considerable economic efficiency" (p. 353).

Once their locations were established, the farms had to be staffed. In Ghana, three of the four agencies that staffed the state farms were units not of the public administration but rather of the governing party! In hiring laborers and staff, the managers of the farms were required to give priority to party activists (Dadson, pp. 26ff). As Dadson states: "A foremost objective . . . as with other public projects, was to extend the control [of the party in power] over the rural population, or to buy the political support of the rural population. Therefore, to begin with, only [party] members were recruited into socialist units" (p. 261).

Lowered prices, such as those for export crops, alienate all rural producers. But the normal outputs of public administration can be selectively offered as compensation and so used to build a coalition supportive of the government in power. We have already noted the passage in 1954 of the bill which froze the cocoa price in Ghana, thereby inspiring the formation of a powerful opposition party, the National Liberation Movement. John Dunn and A. F. Robertson, in their fascinating study of Brong Ahafo, one of the cocoa-growing districts in Ghana, document the government's manipulation of public services there in an effort to cripple opposition to its pricing policy. They note, for example, that "communities in Ahafo which had supported the [government] throughout, like Acherensua . . .

and Kukuom duly received their rewards in the form of major items of government development expenditure, like the secondary school provided by the Ghana Educational Trust" (Dunn and Robertson, p. 327).

Not only the allocation of services but also the structure of public administration in Ghana was manipulated to secure political backing. The government dismissed holders of public office, such as chiefs and headmen, who were supporters of the opposition; it replaced them with those willing to stake their futures on backing the government in power. Also, the local administration was removed from the greater Ashanti Region and given a regional standing in its own right. The result was a virtual region-wide promotion in the status and emoluments of public officeholders, as they moved from district to regional status, or, in the case of the chiefs, from mere chiefly to paramount rank. The cocoa farmers remained primarily concerned with the price of cocoa, and so long as it remained low, they remained disaffected. Nonetheless, through the manipulation of public services, political control was reasserted over the region. As Dunn and Robertson note, the opposition's "hegemony" was "overturned from outside" (p. 341).

Among the public services, it is the agricultural agencies that are of foremost interest to many persons in the countryside, and in building political organizations the governments of Africa manipulate the patronage potential of these agencies. For example, in analyzing the rural base of the governing party of Senegal, Schumacher notes that the party's "strength in the countryside was undoubtedly buttressed by the promotion of party supporters and protégés into key posts in economic and administrative structures directly in touch with the rural population. In addition to cooperative officials, these included produce inspectors [and] secretaries of storage facilities" (p. 16). Schumacher uses this fact to explain the party's rural strength in the face of adverse government pricing policies. Additionally, the cooperatives, which in many countries serve as the local agencies for many government programs, are the bases for governing parties as well. This is true in Kenya, Tanzania, Zambia, and Senegal. In Senegal, efforts to divest the governing party of political control over the cooperative societies led to massive political opposition by rural party leaders and to the downfall of the "technocrats" who advocated this measure (see Schumacher; Donal Cruise O'Brien 1975). The agricultural bureaucracy and its ancillary organizations thus form a fund for political patronage.

By making the attainment of particular benefits—whether a project for a community or a job or promotion for an individual—the substance of rural politics, the governments of Africa have augmented their control over their rural populations. Through the promise of benefits they can secure cooperation; through their conferral, they can reward compliance; and through their withdrawal, they can punish those who protest.

In interviewing a rich cocoa farmer in Ghana in 1978, I asked him why he did not try to organize political support among his colleagues for a rise in product prices. He went to his strongbox and produced a packet of documents: licenses for his vehicles, import permits for spare parts, titles to his real property and improvements, and the articles of incorporation that exempted him from a major portion of his income taxes. "If I tried to organize resistance to the government's policies on farm prices," he said while exhibiting these documents, "I would be called an enemy of the state and I would lose all these." He was a cocoa farmer and we were discussing cocoa prices. The price of Ghanaian cocoa is indeed one of the most politically sensitive topics in African agrarian politics. But in systems where producers operate in markets which are increasingly controlled by public agencies, his point was generally valid.

Through coercion, governments in Africa block the efforts of those who would organize in attempts to achieve structural changes; only the advocacy of minor adjustments is allowed. Moreover, through the conferral of divisible benefits, they make it in the interests of individual rural dwellers to seek limited objectives. Political energies, rather than focusing on the collective standing of the peasantry, focus instead on the securing of particular improvements—subsidized inputs, the location and staffing of production schemes, the allocation of jobs, and the issuance of licenses and permits. Rather than appeals for collective changes, appeals instead focus on incremental benefits. The politics of the pork barrel supplant the politics of class action. Debates over the fundamental configuration of policies remain off the political agenda of the African countryside, and individual rural dwellers come, as a matter of personal self-interest, to abide by public policies that are harmful to agrarian interests as a whole.

Commonalities and Variations:

The Politics of
Agricultural Policy

This book has analyzed the ways in which political power has been used to manipulate the major markets that determine the incomes of farmers in Africa.

Agricultural policies in Africa are characterized by attempts to set prices in markets in a way that is harmful to the interests of most farmers. The economies of Africa are overwhelmingly rural in nature, but the governing elites of Africa seek to industrialize. It is hardly surprising, therefore, that these elites should attempt to extract resources from agriculture and channel them into manufacturing and industry. All nations seeking to industrialize have done this. The African policies are thus notable not as exceptions but as examples of a larger class.

However, in this book we have gone beyond documenting well-established patterns. We have also examined the political processes that underlie efforts to promote the transformation from agriculture to industry. We have seen how governments repress those who would champion the collective interests of agricultural producers; how they give "side payments" to influential members of the rural sector, inducing them to defect from the rural coalition and to ally with those who favor low prices for farm products; and how intervention in markets creates political resources which governments then use to build political organizations. We have also seen how governments engineer a pattern of politics in which the collective standing of agrarian interests is ignored by rural dwellers, while political controversy centers on fights between rural factions for particular benefits. Repression, co-optation, organization, and the promotion of factional conflict: these techniques lie at the heart of the process of rural demobilization. They underlie the political process that secures the triumph of the small fractions of agrarian societies which align with the economic forces that claim ascendancy in the industrial era.

COMMONALITIES

The states of Africa differ in important ways in their policy choices, but behind the differences lie basic similarities in their emergent political economies. It is important that we summarize these similarities.

Fledgling industries locate in the urban areas. Workers and owners, while struggling with each other for their share of industrial profits, possess a common interest in perpetuating policies that increase these profits. They therefore demand policies that shelter and protect these industries. They also demand policies that promise low-cost food.

Because of the public purposes they espouse, African states seek to advance the interests of industry. To secure revenues to promote industry, they therefore seek taxes from agriculture. By maintaining a sheltered industrial order, they generate economic benefits for elites, as well as resources for winning the political backing of influential groups in the urban centers. To safeguard their urban-industrial base, they seek low-cost food. This aim therefore leads them to intervene in markets and to attempt to depress the level of farm prices.

Governments recruit partners in the countryside. Their rural confederates include tenants and managers on state production schemes. They also include elite-level farmers, as well as the more widespread group of progressive farmers who have become dependent on state-sponsored programs of subsidized inputs. These are the rural allies of African regimes—groups that find it privately advantageous to support the governments in power even though they impose disadvantageous agricultural prices. The bureaucracy is another key element in this emergent social order. It spans the markets which governments manipulate. In accordance with public policy, it sets prices within these markets and thereby creates noncompetitive rents. Some of these rents the bureaucrats surrender to the governments in the form of taxes; some they consume themselves; and the remainder they use to build up cadres supportive of governments in power. Through the public management of economic resources, the bureaucracies help to institutionalize a structure of relative advantage—a structure within which they themselves occupy positions of privilege and power.

Scattered around these charter members of the emergent social order in Africa lie the mass of rural producers. They suffer from government attempts to implement an adverse structure of farm prices, and they fail to benefit from the compensatory payments conferred through subsidy programs. Because of official repression, they lack political organizations with which to defend their interests. Furthermore, governments separate the interests of potential rural leaders—the larger farmers—from those of the mass of rural dwellers. As a consequence, instead of pursuing their collective interests, villagers pursue their private interests. They do so by fighting for political favors and by exploiting alternatives left open to them in the private market.

Owners and workers in industrial firms, economic and political elites, privileged farmers and the managers of public bureaucracies—these constitute the development coalition in contemporary Africa. It is they who reap the benefits of the policy choices made in formulating development programs. The costs of these choices are distributed widely, but fall especially hard on the unorganized masses of the farming population.

VARIATIONS

Throughout this book we have noted a series of factors that influence the making of agricultural policy. These factors furnish important sources of variation. They affect the extent and form of government intervention, and they generate differences in the relations between farmers and the state.

Historical Factors

As noted in Chapter One, many African governments inherited marketing boards from their colonial predecessors. While sharing colonial roots, the boards nonetheless differ in their historical origins. Those in East Africa had often been founded by the producers themselves, whereas many of those in West Africa were formed by governments in alliance with trading interests. Though both later became agencies of the independent governments, they differed in their sensitivity to the interests of producers. As shown in the tabulations in Appendix B, the East African producers tend to receive a higher proportion of the world market price for their crops than do their West African counterparts.

A second factor is the social and economic base of the coalition that captured power at the time of self-government. In many instances, the coalition was dominated by urban interests; this was notably the case in Ghana and Zambia, where rural producers formed the backbone of a defeated opposition and thus failed to seize power when independence came. In other instances, rural producers dominated the nationalist movements which succeeded the colonial governments; this was true in the Ivory Coast and, in a more complicated fashion, in Kenya as well. Generally, where urban interests came to power, they have adopted policies more hostile to the interests of farmers; governments dominated by rural producers have intervened less forcibly in the market for outputs and have also provided more favorable subsidies in the markets for inputs.

The Claimants

Other factors that affect the ways in which governments intervene in agricultural markets have to do with the claimants for resources from agriculture. Among the most important claimants are the governments themselves; and, as we have seen, one of the most important factors influencing their behavior is the nature of the "revenue imperative."

The demand for revenue from agriculture, and thus the manner in which governments set prices against it, varies with two factors: the level of governmental commitments to spending programs, and the availability of nonagricultural sources of funds. These help to account for important differences in government pricing policies, and influence the ways in which government treats different segments of agricultural industry.

The importance of the first factor is evident in the changes in policy that took place in many states soon after independence. Self-government brought radically increased commitments to programs of public spending; and as the demand for revenues rose, so did the level of taxation on export agriculture. The importance of the second factor may be seen in the fact that governments with access to nonagricultural sources of funds often impose lighter levels of taxation on export crops. For example, when the oil revenues received by the government of Nigeria increased threefold between 1970 and 1973, one consequence was an upward adjustment of the prices that the marketing boards offered to producers of export crops (see Appendix B).

The way in which a government intervenes in the markets for food crops is strongly influenced by the magnitude of the fiscal resources at its command. Its ability to increase food supplies, either through imports or through the creation of production projects, is limited by the amount of funds it has to spend. Thus the decline of copper revenues in Zambia in the late 1970s led to cutbacks in government programs to support subsidized urban food prices; and conversely, the major programs of farm subsidies so characteristic of production in Nigeria began with the rapid expansion of revenues from oil.

The *types* of fiscal resources available, however, produce different effects. The greater a government's *nonagricultural* revenues, the greater its efforts to increase urban food supplies and subsidize costs to consumers; the higher its subsidies on farm inputs; and the higher the prices it offers to producers of export crops. Greater *agricultural* revenues, on the other hand, promote the same interventions on behalf of the consumers of food crops and in support of farm inputs; but they lead a government to offer lower prices to export crop producers, because the costs of these programs must be covered by taxes on farm products. These important interrelations are shown in Appendix A.

Variations in the strength and structure of the revenue imperative thus help explain variations in the way governments intervene in agricultural markets.

Governments are not the sole claimants of resources from agriculture, however. Processing industries seek raw materials, and industrial workers and owners of firms seek low-cost food. Variations in the strength and behavior of these groups also influence the ways in which governments seek to regulate agricultural markets.

Evidence of the importance of local processing interests is contained in the changes in the prices for export crops offered by the government of Nigeria. As we have seen, beginning in 1973, domestic prices rose as the government secured increasing amounts of its revenues from oil. But for cotton and groundnuts, domestic prices never reached parity with export prices, and an important reason was pressure from domestic processing industries—textile manufacturers in the case of cotton, and producers of vegetable oils in the case of groundnuts (Williams 1976). The larger the fraction of output consumed by domestic processing firms, the greater the pressures on governments to keep down the prices offered to farmers.

Whereas the processors of raw materials look for low-priced cash crops, workers and industrialists look for low-priced food. When urban consumers are poor, expenditures on food consume a higher percentage of their incomes; the lower their incomes, the more they benefit from reductions in food prices. Nations with lower percapita incomes are thus more likely to adopt policies in support of low-priced food. Average incomes in Africa are so uniformly low as to prevent observation of the importance of this factor; but it is of obvious importance in explaining the strength of the pressures upon African governments to appease their urban constituents by providing low-cost food. Moreover, this reasoning has been applied to account for variations in the agricultural policies of different nations of the world in the contemporary era (Schultz 1978), and for variations in the policies of particular nations as they pass through the development process (Hayami; Bates and Rogerson).

Among the basic determinants of the demand for low-priced food by owners of firms are the proportion of their costs represented by wages and the degree to which they can pass higher costs on to consumers. It should be noted that for governments, wages represent a high fraction of their costs, and that few of these costs can be transferred to the consumers of government services. Little wonder, then, that among the industries of Africa, it is the governments themselves which are among the most vocal in calling for low-priced food.

Also important is the degree to which increased food costs are translated into wage claims by workers. In conjunction with what has been argued thus far, we can therefore see why nations with socialist governments adopt agricultural policies that differ little from those of their capitalist neighbors. The stress which socialist regimes place on upgrading the incomes of the rural poor meets a powerful counterforce in the making of farm policy: the higher costs they face from an increase in the price of food. Because they provide a more abundant level of services, they have a greater number of employees; because they own more firms, they are more directly affected by wage claims; and because of their ideological commitments, they are often more responsive to the demands of their workers. An increase in the price of food therefore imposes higher costs on socialist governments, leading them to adopt food-price policies that resemble those of other nations to a greater degree than their official policy positions would lead one to expect.

Characteristics of production

In addition to factors arising on the side of the claimants of resources, important variables operate on the side of the production and marketing of agricultural commodities. These, too, shape the policy choices of governments.

One factor is the nature of the crop. We have noted, for example, that in the case of export crops, governments directly intervene in product markets in attempts to depress prices. For food crops, intervention often takes less direct forms: attempts to change crop prices by altering the availability of supplies or the price of farm inputs. The nature of the crop influences the form of intervention. Export crops often grow in specialized regions. Not all consumers can buy them; they must have access to foreign markets or to expensive processing equipment. And these crops must pass through specific locations—ports and harbors, for example. By contrast, most food crops are grown by most farmers. They may be sold to anyone, for they are often easily processed and directly consumed; and they need not pass through highly specialized marketing channels. It is therefore simply much easier to control the marketing of export crops. As a consequence, in efforts to alter produce prices, governments can more directly intervene in the market for export crops; almost literally, they can seize control of the market. But for food crops, government manipulation must be more indirect; it must take the form of altering supplies and costs of production.

A second major factor is the structure of production. Large farmers, we have seen, often possess close social and political ties with governing elites; and this, we have argued, is of increasing importance in Africa. One consequence is that crops whose production is dominated by large farmers tend to be less heavily taxed. Rice, which is grown by elite farmers in Ghana, is subsidized; Ghanaian cocoa, which is grown largely by small-holders, is heavily taxed. And as seen in Appendix B, the producer price of coffee grown on estates in Kenya lies at over 90 percent of the world price; for coffee grown by smallholders, the producer price stands as less than 66 percent of the world price.

The size distribution of production is also important because it affects the incentives and capacity of farmers to organize in defense of their interests. When the interests of farmers and governments do conflict, large farmers can more readily organize efforts to alter government policy. Groundnuts, for example, are produced in both Nigeria and Senegal. In Senegal, as we have seen, the Marabouts produce one quarter of the crop; in Nigeria, no comparable group dominates groundnut production. Although in both countries the crop is heavily taxed, political protest led by the Marabouts in Senegal resulted in an upward revision in the producer price of groundnuts; in Nigeria there were no organized protests, and the export of groundnuts was banned in an effort to lower the prices paid by local industries. A multitude of small farmers is far more vulnerable to adverse actions by government.

Another factor is the degree of relative advantage that producers hold in the production and marketing of a crop. Ironically, the stronger their relative advantage in production, the weaker their political position. For the stronger their relative advantage, the longer they will persist in growing a crop under conditions of falling prices—the more thoroughly they can be "squeezed," in short, by adverse pricing policies. Where producers hold an advantage in the market for a crop, however, then the obverse is true. If consumers have few alternatives, producers can demand higher prices for their products, and they can more vigorously defend their position in the formulation of pricing policies.

An example of special environmental and ecological conditions that give economic advantages to producers would be the forests of West Africa, which create highly favorable conditions for the production of cocoa. Taking advantage of the inferiority of the producers' second-best alternatives, West African governments have long subjected cocoa to severe levels of taxation. And now that the locational advantage of the producers has been severely eroded by the additional costs imposed by government policies, the advantage of the governments is weakening; as we have seen, producers are increasingly evading the policies of governments by entering the production of alternative crops. And there is increasing evidence that government policies may change.

The effect of consumer alternatives on the pricing policies of governments can perhaps best be illustrated by the different treatment given to export crops and food crops. Export crops are sold in international markets where purchasers can choose between sources of supply. Food crops are sold on domestic markets were alternate sources of supply must come in the form of imports. Because most often imports must come from distant food surplus regions, they represent a costly alternative. All else being equal, the producers of food crops are therefore in a stronger position to secure higher prices than are the producers of export crops—which is another reason for their more favorable treatment by governments.

Differences in historical background, in the characteristics of claimants to resources from agriculture, and in factors associated with the production and marketing of agricultural commodities are all differences that cause variation in the choices made by African governments. Their importance can be illustrated by citing divergent cases.

Palm oil in Southern Nigeria in the 1960s was produced in a nation where marketing boards had been set up by the government in association with merchant interests. Government revenues derived from export agriculture, and popular demands for government services were strong; local processors consumed a growing share of the industry's output; farmers had few alternative cash crops; and production was in the hands of small-scale, village-level farmers. The industry was subject to a high level of taxation. Only when farmers began to abandon the production of palm oil for other crops, and when the government found different sources of revenue, did the government relent and offer higher prices for the crop.

The production of wheat in Kenya offers a striking contrast. Historically, the marketing board for wheat had been set up by the producers themselves and prosperous indigenous farmers had played a major role in the nationalist movement which seized power in the post-independence period. The government derived a relatively small portion of its revenues from agriculture; farmers had attractive alternatives to the production of wheat; consumers had a strong preference for wheat products, and alternative sources of supply lay in distant foreign markets. Wheat production was dominated by a relatively small number of very large farmers; and elite-level figures had direct financial interests in wheat farming. The result was a set of policies providing favorable prices for wheat products and extensive subsidies for farm inputs.

A major objective of this study has been to identify factors that help account for the ways in which governments intervene in agricultural markets. We have located three types of factors. And, as we have seen, they help to account for differences in public policies and for variations in the relations between the farmer and the state in Africa.

ALTERNATIVE FUTURES

The agricultural policies of the nations of Africa confer benefits on highly concentrated and organized groupings. They spread costs over the masses of the unorganized. They have helped to evoke the self-interested assent of powerful interests to the formation of a new political order, and have provoked little organized resistance. In this way, they have helped to generate a political equilibrium. But in the longer run, the costs inflicted by these policies are being passed on to members of the policy-making coalition, and the configurations that were once in equilibrium are now becoming politically unstable.

Among those excluded from the immediate rewards of the new political order are the mass of farmers. For the benefit of others, they are subjected to policies that violate their interests. But the effects of these policies are increasingly harmful to everyone. Reducing the incentives to grow food leads to reduced food production; the result is higher food prices and waves of discontent in the urban centers. The coups and counter-coups that have recently swept West Africa owe their origin, as we have noted, in part to discontent over higher food prices. And they show how policies that have been designed to serve the interests of powerful groups impose costs which in the long run affect everyone, thereby undercutting the positions of advantage they have helped to create and disrupting the political order.

Reducing the incentives to produce export crops is also proving politically costly. The result of adverse incentives has in some cases been a measurable decline in the production of exports, with a resultant loss of public revenues and foreign exchange. Pressed for revenues, governments have to cut back on politically popular programs—food subsidies to appease the urban consumer, input subsidies to tame the rural elite, or schools, roads, and clinics to reward communities that have kept political good faith.

As governments throughout the developing world have learned, financial retrenchment generates political crisis. In the face of fewer revenues, governments do have an alternative: they can spend beyond their means. But this choice, too, fuels political disaffection by strengthening the forces of inflation. Shortages of foreign exchange have meant decreased imports, and shortages of imports provoke cutbacks in production by industry and in services by governments. The result, once again, is growing political discontent. Too often, then, policies adopted to extract revenues from export agriculture have led to an increasing scarcity of the resources used to underpin the political order.

The costs that were inflicted on farmers are thus increasingly being transferred to those who initially benefited from the selection of agricultural policies. The basis of the equilibrium erodes. Incentives are thereby created for altering policy choices. Here the question arises: Are there politically workable terms under which the farmers could be admitted to the governing coalition, and thereby receive support for policies more favorable to their interests?

The following scenario might work for food producers. In nations with a major extractive industry, a coalition of urban industry and food producers could unite against the extractive industry. Food producers could use the tax revenues generated by the extractive industry to subsidize the cost of inputs. Government-owned railways could charge high prices for carrying the freight of the extractive industry in order to subsidize the rail charges for agricultural products, for example. Food producers could also use public revenues to support high producer prices; they can use them, for example, to finance storage and exports, thereby withdrawing excess supplies from the domestic market and protecting a high level of domestic prices. Industry, for its part, could seek an overvalued currency and tariff protection. An overvalued currency would facilitate imports of capital, paid for with foreign exchange earned by the extractive industry; tariff protection would shield local markets from foreign competition and also offset the increase in wages resulting from higher-priced food. Alternatively, industry could secure the use of government revenues to subsidize urban food prices, thereby offsetting the threat of higher wage demands.

Coalitions such as these have in fact formed in Africa, as in the settler territories of southern Africa. They may now begin to form in areas where discoveries of oil or uranium may bring radical increases in the wealth of certain black African nations. And they become more likely as greater numbers of politically influential persons begin to engage in domestic food production.

In the case of export crops, the scenario differs. The most obvious tradeoff that would bring the producers of export crops into the dominant coalition would be one in which measures to reduce the overvaluation of the national currency are exchanged for measures to reduce the cost of industrial labor. As we have seen, shifting to an equilibrium exchange rate instead of overvaluing the local currency would favor the interests of exporters. Such a measure, all other things being equal, would greatly increase the incomes of the producers of export crops. But it would do so at the expense of others, such as industry, who must import from abroad. To secure a coalition between export agriculture and industry, compensation must be offered for this loss of advantage. In part, such compensation can take the form of higher levels of prosperity among rural producers, for this expands the size of the markets for manufactured products. But their prosperity also poses a threat, for it leads to competitive bids for labor by the more prosperous countryside. The most direct form of compensation that can be offered, therefore, will consist of measures designed to cheapen the costs of labor: repressive labor laws which limit collective action by workers, and laws which give foreign workers relatively open access to the domestic labor market. Export-oriented policies on the one hand, and policies in support of cheap labor on the other, thus form the basis for a coalition that incorporates the interests of export agriculture.

Such a bargain has in fact been struck on one nation—the Ivory Coast. As students of that nation have long appreciated, an important basis for its policies is that members of the elite derive large portions of their incomes from the production of export crops. This fact assumes particular importance in the context of our argument. For just as overvaluation cheapens the costs of imported equipment for firms, it also cheapens the costs of imported goods for consumers. The elite, being rich and having "modern" tastes, consume a disproportionate share of imports, and therefore can be expected to resist devaluation of the local currency. Insofar as the elite draws a major portion of its income from exports, however, the benefits of the change in commercial policy compensate for its loss in foreign purchasing power. As less sacrifice is required from the politically influential, it is politically easier to secure this shift in commercial policy. Here again we can see the importance of the rural origins of the political elite for the making of public policy.

These scenarios portray politically workable grounds for incorporating the interests of the two kinds of producers into the policy-making coalitions of Africa. In both instances, exceptional conditions give political impetus to the favoring of their interests. In most cases, these conditions will be absent. As a consequence powerful actors—revenue-starved governments, price-conscious consumers, profit-seeking industries, and dependent farmers—will persist in seeking their individual, short-run, best interests, and they will continue to adhere to policy choices that are harmful to farmers and collectively deleterious as well. Producers will prosper, then, only insofar as they successfully evade the prescriptions of their governments.

Alternatively, in response to the erosion of advantages engendered by shortfalls in production, the dominant interests may be persuaded to forsake the pursuit of unilateral short-run advantage, and instead to employ strategies that evoke cooperation by sharing joint gains. In the face of mounting evidence of the failure of present policies, people may come to believe that short-run price increases for farmers may in the longer run lead to more abundant supplies and less costly food; or that decreases in tax rates may lead to greater revenues as a result of increased production; or that positive incentives for greater production may lead to greater production and lower prices and leave only the most efficient farms in production, thereby accelerating a shift of resources from agriculture to industry. The growth of an awareness that present measures offer few incentives for farmers to play a positive role in the great transformation may thus provide a foundation for attempts to reform the agricultural policies of the nations of Africa.

Interrelations Between Food Supply, Demand, and Prices

The object is to reduce the price of food, say, from P_0 to P_1. To secure the lower price, the government attempts to increase the quantity supplied from Q_{s_0} to Q_{s*} (Graph 1).

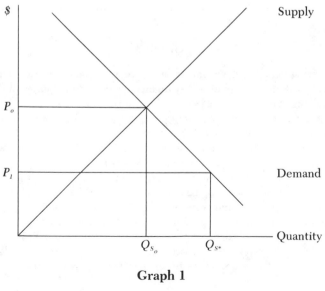

Graph 1

If the government can secure the food abroad at P_1, the cost of importing the food to increase supplies to Q_{s^*} can then be represented as the shaded area in Graph 2. At P_1 local producers will supply Q_{s_1} and the government, through imports, makes up the difference between the amount produced locally (Q_{s_1}) and the amount needed to support the price at P_1 (Q_{s^*}).

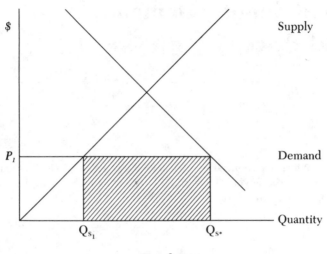

Graph 2

Alternatively, the government can increase supplies by supporting the costs of local production, either by financing food production schemes or by subsidizing the production costs of local producers. The cost of securing increasing local production from Q_{s_0} to Q_{s^*}, thereby depressing the price to P_1, is the shaded area in Graph 3.

Whichever tactic is chosen, governments in Africa tend to pass these costs of increasing food supplies back onto a portion of the farming population. In particular, they tend to pass them on to the export producer who generates revenue and foreign exchange. Perhaps the most vivid example is provided by Sierra Leone, where the government compelled the marketing board—which handles agricultural exports and acts as a taxing agent for the government—to take over management of the national rice corporation—which is responsible for provisioning the urban areas with low-cost rice. The board now provides the finances for importing rice and for pro-

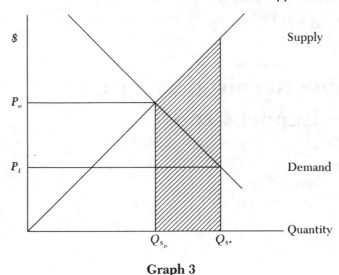

Graph 3

moting domestic production by subsidizing the costs of farm inputs (*African Business*, 1980).

In this way the need for revenues to provide low-priced food for consumers, and low-cost inputs for food producers, leads to the purposeful reduction in the price of farm exports. This is not an untypical configuration of policy choices.

APPENDIX B

Value Received by Farmers for Export Crops

The following table suggests the extent to which farmers receive the value of the crops which they produce for export. For a series of crops, the table reports the percent of the sales realization which the farmers actually receive. It also lists the sources from which the information was taken.

In some cases, which I have marked (p), the measure is based on prices. It is the ratio of the price received by the producer to the price that prevailed on the world market. In each instance, I have used sources that employed the f.o.b. price at the major national port as a measure of the world price. In other cases, which I have marked (i), the measure is calculated in terms of incomes. It is then the ratio of the total value of the farmers' earnings from the sale of the crop to the reported total value realized from the sale of the crop on the international market.

Value Received by Farmers
for Crops They Produce for Export
(by crop and country)

(i) = Percent of income from the sale of crop obtained by producer.
(p) = Price paid to producer as percent of international (f.o.b.) price.
(b) = No international sales of cotton.

Crop	Country	Year	*Percent of sales realization by farmers*	Source
Cotton	Kenya	1970–1971	82 (i)	Gray 1977
		1971–1972	66 (i)	
		1975–1976	48 (i)	
	Sudan	1961–1962	44 (i)	ILO 1976
		1971–1972	49 (i)	
	Nigeria	1950–1951	16 (p)	Onitiri-Olatunbosun 1974
		1951–1952	17 (p)	
		1952–1953	16 (p)	
		1953–1954	17 (p)	
		1954–1955	20 (p)	
		1955–1956	20 (p)	
		1956–1957	20 (p)	
		1957–1958	22 (p)	
		1958–1959	24 (p)	
		1959–1960	28 (p)	
		1960–1961	25 (p)	
		1961–1962	20 (p)	
		1962–1963	18 (p)	
		1963–1964	19 (p)	
		1964–1965	21 (p)	IBRD 1978
		1965–1966	21 (p)	
		1966–1967	23 (p)	
		1967–1968	24 (p)	
		1968–1969	27 (p)	
		1969–1970	32 (p)	
		1970–1971	36 (p)	
		1971–1972	46 (p)	

Value Received by Farmers
for Crops They Produce for Export
(continued)

		1972–1973	43 (p)	
		1973–1974	— (b)	
		1974–1975	— (b)	
		1975–1976	— (b)	
		1976–1977	95 (p)	
	Tanzania	1966–1967	65 (p)	Republic of Tanzania 1976
		1967–1968	58 (p)	
		1968–1969	59 (p)	
		1969–1970	71 (p)	
		1970–1971	64 (p)	
		1971–1972	54 (p)	
		1972–1973	55 (p)	
		1973–1974	36 (p)	
		1974–1975	41 (p)	
	Uganda	1954	70 (i)	Jamal 1976
		1955	75 (i)	
		1956	77 (i)	
		1957	76 (i)	
		1958	100 (i)	
		1959	101 (i)	
		1960	75 (i)	
		1954–1960	80 (i)	
Cocoa	Nigeria	1947–1948	65 (p)	Onitiri-Olatunbosun 1974
		1948–1949	61 (p)	
		1949–1950	71 (p)	
		1950–1951	63 (p)	
		1951–1952	66 (p)	
		1952–1953	68 (p)	
		1953–1954	70 (p)	
		1954–1955	49 (p)	
		1955–1956	66 (p)	
		1956–1957	71 (p)	
		1957–1958	76 (p)	
		1958–1959	48 (p)	

Value Received by Farmers
for Crops They Produce for Export
(*continued*)

	1959–1960	58 (p)	
	1960–1961	62 (p)	
	1961–1962	52 (p)	
	1962-1963	59 (p)	
	1963–1964	57 (p)	
	1964–1965	89 (p)	
	1965–1966	51 (p)	
	1966–1967	45 (p)	
	1967–1968	43 (p)	
	1968–1969	38 (p)	
	1964–1965	89 (p)	IBRD 1978
	1965–1966	39 (p)	
	1966–1967	46 (p)	
	1967–1968	38 (p)	
	1968–1969	34 (p)	
	1969–1970	45 (p)	
	1970–1971	50 (p)	
	1971–1972	62 (p)	
	1972–1973	58 (p)	
	1973–1974	50 (p)	
	1974–1975	63 (p)	
	1975–1976	72 (p)	
	1976–1977	66 (p)	
Ghana	1947	56 (p)	Bateman 1965
	1948	38 (p)	
	1949	89 (p)	
	1950	41 (p)	
	1951	49 (p)	
	1952	61 (p)	
	1953	55 (p)	
	1954	34 (p)	
	1955	40 (p)	
	1956	67 (p)	
	1957	74 (p)	
	1958	42 (p)	

Value Received by Farmers
for Crops They Produce for Export
(*continued*)

		1959	48 (p)	
		1960	51 (p)	
		1961	66 (p)	
		1962	65 (p)	
		1962–1963	62 (p)	Beckman 1976
		1963–1964	57 (p)	
		1964–1965	60 (p)	
		1947–1948	37 (i)	
		1948–1949	90 (i)	
		1949–1950	46 (i)	
		1950–1951	49 (i)	
		1951–1952	61 (i)	
		1952–1953	56 (i)	
		1953–1954	38 (i)	
		1954–1955	38 (i)	
		1955–1956	65 (i)	
		1956–1957	78 (i)	
		1957–1958	44 (i)	
		1958–1959	48 (i)	
		1959–1960	51 (i)	
		1960–1961	68 (i)	
		1961–1962	60 (i)	
		1962–1963	62 (i)	
		1963–1964	55 (i)	
Coffee	Kenya, small-holders	1970–1971	63 (p)	ILO 1977
		1971–1972	62 (p)	
		1972–1973	62 (p)	
		1973–1974	61 (p)	
		1974–1975	63 (p)	
		1975–1976	64 (p)	
	Kenya, estates	1970–1971	92 (p)	
		1971–1972	91 (p)	
		1972–1973	90 (p)	
		1973–1974	90 (p)	

Value Received by Farmers
for Crops They Produce for Export
(continued)

		1974–1975	93 (p)	
		1975–1976	93 (p)	
	Tanzania*	1971–1972	75 (p)	
		1972–1973	69 (p)	
		1973–1974	57 (p)	
		1974–1975	66 (p)	
		1975–1976	58 (p)	
		1976–1977	46 (p)	Tanzania 1977
	Uganda	1954	76 (i)	Jamal 1976
		1955	116 (i)	
		1956	81 (i)	
		1957	77 (i)	
		1958	72 (i)	
		1959	86 (i)	
		1960	127 (i)	
		1954–1960	90 (i)	
Pyrethrum	Kenya	1970–1971	75 (i)	Gray 1977
		1971–1972	70 (i)	
		1972–1973	67 (i)	
		1973–1974	62 (i)	
		1974–1975	77 (i)	
		1975–1976	66 (i)	
Wattle Bark	Kenya	1970–1971	39 (i)	Gray 1977
		1971–1972	38 (i)	
		1972–1973	35 (i)	
		1973–1974	33 (i)	
		1974–1975	28 (i)	
		1975–1976	28 (i)	
Groundnuts	Nigeria	1947–1948	64 (p)	Onitiri-Olatunbosun 1974
		1948–1949	48 (p)	

*Price paid to smallholders as percent of auction price; mild Arabica coffee.

Value Received by Farmers
for Crops They Produce for Export
(*continued*)

	1949–1950	42 (p)	
	1950–1951	44 (p)	
	1951–1952	55 (p)	
	1952–1953	42 (p)	
	1953–1954	48 (p)	
	1954–1955	51 (p)	
	1955–1956	61 (p)	
	1956–1957	52 (p)	
	1957–1958	56 (p)	
	1958–1959	65 (p)	
	1959–1960	66 (p)	
	1960–1961	54 (p)	
	1961–1962	58 (p)	
	1962–1963	51 (p)	
	1963–1964	48 (p)	
	1964–1965	48 (p)	IBRD 1978
	1965–1966	47 (p)	
	1966–1967	50 (p)	
	1967–1968	46 (p)	
	1968–1969	41 (p)	
	1969–1970	40 (p)	
	1970–1971	37 (p)	
	1971–1972	37 (p)	
	1972–1973	35 (p)	
	1973–1974	42 (p)	
	1974–1975	50 (p)	
	1975–1976	83 (p)	
	1976–1977	120 (p)	
Senegal	1962–1963	45 (p)	IBRD 1974
	1963–1964	45 (p)	
	1964–1965	45 (p)	
	1965–1966	48 (p)	
	1966–1967	46 (p)	
	1967–1968	47 (p)	
	1968–1969	46 (p)	

**Value Received by Farmers
for Crops They Produce for Export**
(*continued*)

		1969–1970	36 (p)	
		1970–1971	32 (p)	
		1971–1972	40 (p)	
		1972–1973	30 (p)	
		1962–1963	65 (i)	
		1963–1964	65 (i)	
		1964–1965	65 (i)	
		1965–1966	69 (i)	
		1966–1967	67 (i)	
		1967–1968	67 (i)	
		1968–1969	66 (i)	
		1969–1970	52 (i)	
		1970–1971	46 (i)	
		1971–1972	57 (i)	
		1972–1973	43 (i)	
Palm Oil	Nigeria	1947–1948	38 (p)	Onitiri-Olatunbosun 1974
		1948–1949	54 (p)	
		1949–1950	61 (p)	
		1950–1951	61 (p)	
		1951–1952	64 (p)	
		1952–1953	60 (p)	
		1953–1954	117 (p)	
		1954–1955	87 (p)	
		1955–1956	81 (p)	
		1956–1957	62 (p)	
		1957–1958	60 (p)	
		1958–1959	67 (p)	
		1959–1960	57 (p)	
		1960–1961	63 (p)	
		1961–1962	59 (p)	
		1962–1963	53 (p)	
		1963–1964	54 (p)	
		1964–1965	48 (p)	IBRD 1978
		1965–1966	45 (p)	

Value Received by Farmers
for Crops They Produce for Export
(*continued*)

	1966–1967	54 (p)	
	1967–1968	55 (p)	
	1968–1969	91 (p)	
	1969–1970	91 (p)	
	1970–1971	49 (p)	
	1971–1972	56 (p)	
	1972–1977	Foreign exports ceased	
Palm Kernel Nigeria	1947–1948	36 (p)	Onitiri-Olatunbosun 1974
	1948–1949	60 (p)	
	1949–1950	58 (p)	
	1950–1951	64 (p)	
	1951–1952	55 (p)	
	1952–1953	59 (p)	
	1953–1954	62 (p)	
	1954–1955	69 (p)	
	1955–1956	68 (p)	
	1956–1957	66 (p)	
	1957–1958	68 (p)	
	1958–1959	63 (p)	
	1959–1960	48 (p)	
	1960–1961	47 (p)	
	1961–1962	60 (p)	
	1962–1963	54 (p)	
	1963–1964	48 (p)	
	1964–1965	46 (p)	IBRD 1978
	1965–1966	45 (p)	
	1966–1967	51 (p)	
	1967–1968	48 (p)	
	1968–1969	45 (p)	
	1969–1970	51 (p)	
	1970–1971	52 (p)	IBRD 1978
	1971–1972	74 (p)	

Value Received by Farmers
for Crops They Produce for Export
(continued)

1972–1973	41 (p)
1973–1974	40 (p)
1974–1975	52 (p)
1975–1976	150 (p)
1976–1977	130 (p)

Bibliography

Adesimi, A. A.
1970. "An Econometric Study of Air-Cured Tobacco Supply in Western Nigeria, 1945–1964." *The Nigerian Journal of Economic and Social Studies* 12, 3:315–322.

Alavi, Hamza
1972. "The State in Post-Colonial Societies—Pakistan and Bangladesh." *New Left Review* 74:59–81.

Alibaruho, George
1974. "African Farmer Response to Price: A Survey of Empirical Evidence." Working Paper No. 177. Institute for Development Studies. University of Nairobi.

Amin, Samir
1973. *Neo-Colonialism in West Africa*. New York: Monthly Review Press.

Amey, Alan B. and David K. Leonard
1978. "Public Policy, Class and Inequality in Kenya and Tanzania." Mimeographed. Berkeley, Calif.

Anthony, Kenneth R. M., Bruce F. Johnston, William O. Jones, and Victor C. Uchendu
1979. *Agricultural Change in Tropical Africa*. Ithaca, N.Y.: Cornell University Press.

148 Bibliography

Armah, Matthew Eric
1977. "The Present and Potential Use of Credit by Small Scale Farmers in the Agona Swedru District." Dissertation. Department of Agricultural Economics, University of Ghana.

Arrighi, Giovanni
1973. "International Corporations, Labor Aristocracies, and Economic Development in Tropical Africa." In *Essays on the Political Economy of Africa*, edited by Giovanni Arrighi and John Saul. New York and London: Monthly Review Press.

Askarai, Hossein, and John Thomas Cummings
1976. *Agricultural Supply Response: A Survey of the Econometric Evidence*. New York: Praeger Publishers.

Austin, Dennis
1970. *Politics in Ghana, 1946–1960*. London: Oxford University Press.

Awolowo, Obafemi
1960. *Awo: The Autobiography of Chief Obafemi Awolowo*. Cambridge: The University Press.

Balassa, Bela
1978. "Comparative Advantage and Perspectives for Economic Integration in Western Africa." Paper prepared for the Colloquium on the Integration of Western Africa, Dakar, Senegal. Mimeographed.

Bale, Malcolm D., and Ernst Lutz
1979. "Price Distortions in Agriculture and Their Effects: An International Comparison." World Bank Staff Working Paper No. 359. Washington, D.C.: World Bank.

Barkan, Joel D.
1975. "Bringing Home the Pork: Legislator Behavior, Rural Development and Political Change in East Africa." Occasional Paper No. 9. Iowa City: Comparative Legislative Research Center, University of Iowa.

1976. "Comment: Further Reassessment of 'Conventional Wisdom': Political Knowledge and Voting Behavior in Rural Kenya." *American Political Science Review* 70, 2:452–455.

1979. "Comparing Politics and Public Policy in Kenya and Tanzania." In *Politics and Public Policy in Kenya and Tanzania*, edited by Joel D. Barkan with John J. Okumu, pp. 3–40. New York: Praeger Publishers.

Barkan, Joel D., and John J. Okumu
1974. "Political Linkage in Kenya: Citizens, Local Elites, and Legislators." Paper prepared for the Seventieth Annual Meeting of the

American Political Science Association, Chicago. Mimeographed.

Barnett, Tony
1977. *The Gezira Scheme: An Illusion of Development.* London: Frank Cass.

Barnum, H. N., and R. H. Sabot
1977. "Education, Employment Probabilities and Rural-Urban Migration in Tanzania." *Oxford Bulletin of Economics and Statistics* 39, No. 2:109–126.

Bateman, Merril J.
1965. "Aggregate and Regional Supply Functions for Ghanaian Cocoa, 1946–1962." *Journal of Farm Economics* 47, 2:384–401.

Bates, Robert H.
1976. *Rural Responses to Industrialization: A Study of Village Zambia.* New Haven and London: Yale University Press.

1980. "Pressure Groups, Public Policy and Agricultural Development: A Study of Divergent Outcomes." In *Agricultural Development in Africa: Issues of Public Policy,* edited by Robert H. Bates and Michael F. Lofchie. New York: Praeger Publishers.

Bates, Robert H., and William P. Rogerson
Forthcoming (1981). "Agriculture in Development: A Coalitional Analysis." *Public Choice.*

Bauer, P. T.
1964. *West African Trade.* London: Routledge and Kegan Paul.

Baylies, Carolyn
1978. "The State and Class Formation in Zambia." 2 vols. Ph.D. Dissertation, University of Michigan.

1979. "The Emergence of Indigenous Capitalist Agriculture: The Case of Southern Province, Zambia." *Rural Africana,* Nos. 4–5, 65–82.

Beals, Ralph E., Mildred B. Levy, and Leon Moses
1967. "Rationality and Migration in Ghana." *Review of Economics and Statistics* 49, 4:480–486.

Beckman, Björn
1976. *Organizing the Farmers: Cocoa Politics and National Development in Ghana.* New York: Holmes and Meier.

Berg, Elliot J.
1964. "Real Income Trends in West Africa, 1939–1960." In *Economic Transition in Africa,* edited by Melvill J. Herskovits and Michael Harwitz, pp. 199–238. Evanston, Ill.: Northwestern University Press.

Bergsman, Joel
 1979. "Growth and Equity in Semi-Industrialized Countries." World
        Bank Staff Working Paper No. 351. Washington, D.C.: World
        Bank.
Bottrall, A. F.
 1976. "Financing Small Farmers: A Range of Strategies." In *Policy and
        Practice in Rural Development*, edited by Guy Hunter, A. H.
        Bunting, and Anthony Bottrall, pp. 355–370. London: Croom
        Helm.
Brigg, Pamela
 1971. "A Survey of Case Studies on Migration to Urban Areas." Wash-
        ington, D.C.: International Bank for Reconstruction and Devel-
        opment.
Buijtenhuijs, Robert
 1973. *Mau Mau: Twenty Years After*. The Hague: Mouton and Com-
        pany.
Byerlee, Derek
 1972. "Research on Migration in Africa." African Rural Employment Pa-
        per No. 2. East Lansing: Department of Agricultural Economics,
        Michigan State University.
Callaghy, Thomas M.
 1976. "Implementation of Socialist Strategies of Development in Africa:
        State Power, Conflict and Uncertainty." Paper presented to the
        1976 Annual Meeting of the American Political Science Associa-
        tion, Chicago. Mimeographed.
Campbell, Bonnie
 1974. "The Social, Political and Economic Consequences of French Pri-
        vate Investment in the Ivory Coast, 1960–1970. A Case Study of
        Cotton and Textile Production." Ph.D. Dissertation, University
        of Sussex.
 1978. "The Ivory Coast." In *West African States: Failure and Promise*,
        edited by John Dunn, pp. 66–116. Cambridge: Cambridge Uni-
        versity Press.
Center for Policy Alternatives
 1974. *A Framework for Evaluating Long-Term Strategies for the Devel-
        opment of the Sahel-Sudan Region*, Annex 1. Economic Consid-
        erations for Long-Term Development. Cambridge: Massachusetts
        Institute of Technology.
Clark, W. Edmund
 1978. *Socialist Development and Public Investment in Tanzania,
        1964–1973*. Toronto: University of Toronto Press.

Cliffe, Lionel, ed.
1967. *One Party Democracy: The 1965 Tanzania General Elections.* Nairobi: East African Publishing House.
Club du Sahel
1977. *Marketing, Price Policy and Storage of Food Grains in the Sahel, A Survey.* (Club du Sahel; Working Group on Marketing, Price Policy, and Storage.) Volume 1. Ann Arbor: University of Michigan, Center for Research on Economic Development.
Cohen, John, and Dov Weintraub
1975. *Land and Peasants in Imperial Ethiopia: The Social Background to Revolution.* Assen: Van Gorcum.
Cohen, Michael A.
1974. *Urban Policy and Political Conflict in Africa.* Chicago and London: University of Chicago Press.
Cohen, Robin
1974. *Labour and Politics in Nigeria, 1945–1971.* New York: Africana Publishing Company.
Coleman, James Smoot
1958. *Nigeria: Background to Nationalism.* Berkeley: University of California Press.
Cruise O'Brien, Donal B.
1971. *The Mourides of Senegal: The Political and Economic Organization of an Islamic Brotherhood.* Oxford: Clarendon Press.
1975. *Saints and Politicians: Essays in the Organization of a Senegalese Peasant Society.* African Studies Series No. 15. London and New York: Cambridge University Press.
1979. "Ruling Class and Peasantry in Senegal, 1960–1976." In *Political Economy of Underdevelopment: Dependence in Senegal,* edited by Rita Cruise O'Brien, pp. 209–227, Sage Series on African Modernization and Development, 3. Beverly Hills, Calif.: Sage Publications.
Cruise O'Brien, Rita
1979. *The Political Economy of Underdevelopment: Dependence in Senegal.* Beverly Hills, Calif.: Sage Publications.
Dadson, John Alfred
1970. "Socialized Agriculture in Ghana, 1962–1965." Ph.D. Dissertation, Department of Economics, Harvard University.
DeWilde, John C.
1977. "Price Incentives and African Agricultural Development." Paper presented to the Spring Seminar of the African Studies Program of the University of California at Los Angeles. Mimeographed.

152   Bibliography

Dixon-Fyle, Mac
1974.   "The Genesis and Development of African Protest on the Tonga
        Plateau, 1900–53." Seminar Paper No. 13. Department of His-
        tory. University of Zambia. Mimeographed.
Dodge, Doris Jansen
1977.   Agricultural Policy and Performance in Zambia. Berkeley, Calif.:
        Institute of International Studies.
Dunn, John
1975.   "Politics in Asunafo." In Politicians and Soldiers in Ghana,
        1966–1972, edited by Dennis Austin and Robert Luckham, pp.
        164–213. London: Frank Cass.
1976.   "The Eligible and the Elect: Arminian Thoughts on the Social
        Predestination of Ahafo Leaders." In The Making of Politicians:
        Studies from Africa and Asia, edited by W. H. Morris-Jones, pp.
        49–65. London: The Athlone Press.
Dunn, John, and A. F. Robertson
1973.   Dependence and Opportunity: Political Change in Ahafo. Cam-
        bridge: Cambridge University Press.
Eglin, Richard
1978.   "The Oligopolistic Structure and Competition Characteristics of
        Direct Foreign Investment in Kenya's Manufacturing Sector." In
        Readings on the Multinational Corporation in Kenya, edited by
        Raphael Kaplinsky, pp. 96–133. Nairobi: Oxford University.
Ekhomu, David Onaburekhale
1978.   "National Food Policies and Bureaucracies in Nigeria: Legitima-
        tion, Implementation, and Evaluation." Paper presented at the
        African Studies Association Convention, Baltimore. Mimeographed.
Ewusi, Kodwo
1977.   Economic Inequality in Ghana. Legon, Ghana: Institute of Statis-
        tical, Social and Economic Research, University of Ghana.
1980.   Planning for the Neglected Rural Poor. In press.
Fajana, Olufemi
1977.   "Import Licensing in Nigeria." Development and Change 8:
        509–522.
Fanon, Frantz
1963.   The Wretched of the Earth. New York: Grove Press.
Fitch, Robert Beck, and Mary Oppenheimer
1966.   Ghana: End of an Illusion. New York: Monthly Review Press.
Food and Agricultural Organization
1978a.   FAO Trade Yearbook 1977.

1978b. *FAO Production Yearbook 1977.*

Friedland, William H.

1974. "African Trade Union Studies: Analysis of Two Decades." *Cahiers D'Etudes Africaines* 14, 55:575–589.

Gerhart, John

1975. *The Diffusion of Hybrid Maize in Western Kenya.* Mexico City: International Maize and Wheat Improvement Center.

Ghana

1956. *Report of the Commission of Enquiry into the Affairs of the Cocoa Purchasing Company* (Jibowu Commission).

1965. *Report of the Commission of Enquiry into Trade Malpractices in Ghana* (Abraham Commission).

1967a. *Government Statement on the Report of the Committee Appointed to Enquire into the Local Purchasing of Cocoa* (White Paper No. 3). Accra: Government Printer.

1967b. *Report of the Commission of Enquiry into the Local Purchasing of Cocoa.*

1967c. *Report of the Jiagge Commission.*

1967d. *White Paper on the Report of the Commission of Enquiry into Alleged Irregularities and Malpractices in Connection with the Grant of Import Licenses* (W.P. No. 4). Accra: Government Printer.

1968. *Report of the Commission Appointed to Enquire into the Functions, Operation, and Administration of the Workers' Brigade* (Kom Commission).

1968–69. *Report of the Jiagge Commission—Assets of Specified Persons.* Volumes I–III.

1972. *Outline of Government Economic Policy.* Accra: Government Printer.

1971. Central Bureau of Statistics. *Industrial Statistics 1969.*

1966. Office of the President. *Report of the Commission of Enquiry into Trade Malpractices in Ghana.*

1973 and 1975. Ministry of Agriculture. *Report on Current Agricultural Statistics 1972 and 1974.*

1978. Ministry of Cocoa Affairs. *Ashanti Cocoa Project, Annual Report 1976–77.*

1972. Ministry of Finance and Planning. *Budget Statement for March to June 1972,* by Col. I. K. Acheampong.

Girdner, Janet, and Victor Olorunsula

1978. "National Food Policies and Organizations in Ghana: Legitimation, Implementation, and Evaluation." Paper presented to the Annual Meeting of the American Political Science Association, New York. Mimeographed.

Godfrey, Martin, and Steven Langdon
1976. "Partners in Underdevelopment? The Translation Thesis in a Kenyan Context." *Journal of Commonwealth and Comparative Politics* 14:42–63.

Gordon, J.
1970. "State Farms in Ghana." In *International Seminar on Change in Agriculture*, edited by A. H. Bunting, pp. 577–583. New York: Praeger.

Gray, Clive S.
1977. "Costs, Prices and Market Structure in Kenya." Nairobi. Mimeographed.

Grayson, Leslie E.
1979. *Managing the Economic Development of Ghana*. Charlottesville: Printing Services of the University of Virginia.

Griffin, Keith
1972. *The Green Revolution: An Economic Analysis*. Geneva: United Nations Research Institute.

Gutkind, Peter C. W., Robin Cohen, and Jean Copans, eds.
1978. *African Labor History*. Sage Series on African Modernization and Development, Volume 2. Beverly Hills, Calif.: Sage Publications.

Hayami, Yujiro
1975. "Japan's Rice Policy in Historical Perspective." *Food Research Studies* 14, No. 4:359–380.

Hazelwood, Arthur
1953–1954. "Colonial External Finance Since the War." *The Review of Economic Studies* 21, No. 54:31–52.

Helleiner, G. K.
1964. "The Eastern Nigeria Development Corporation: A Study in Sources and Uses of Public Development Funds, 1949–1962." *Nigerian Journal of Economics and Social Studies* 6, 1:98–123.
1966. *Peasant Agriculture, Government, and Economic Growth in Nigeria*. Homewood, Ill.: Richard D. Irwin, Inc.
1972. "Agricultural Export Pricing Strategy in Tanzania." *East African Journal of Rural Development* 1:1–17.

Hill, Frances
1977. "Experiments with a Public Sector Peasantry." *African Studies Review* 20, No. 3:25–41.

Hirschman, Albert O.
1968. "The Political Economy of Import-Substituting Industrialization in Latin America." *The Quarterly Journal of Economics* 82, No. 1:1–32.
1979. *Exit, Voice, and Loyalty.* Cambridge, Mass.: Harvard University Press.
1979. "The Turn to Authoritarianism in Latin America and the Search for its Economic Determinants." In *The New Authoritarianism in Latin America,* edited by David Collier. Princeton, N.J.: Princeton University Press.
House, William J.
1971. "Market Structure and Industry Performance: The Case of Kenya." Discussion Paper No. 116. Department of Economics, University of Nairobi. Mimeographed.
Hunt, Diana
1975. *Credit for Agricultural Development, A Case Study of Uganda.* Nairobi: East African Publishing House.
Huntington, H.
1974. "An Empirical Study of Ethnic Linkages in Kenyan Rural-Urban Migration." Ph.D. Dissertation, State University of New York, Binghamton.
Hyden, Goran
1980a. *Beyond Ujamaa: Underdevelopment and an Uncaptured Peasantry.* Berkeley and Los Angeles: University of California Press.
1980b. "The Resilience of the Peasant Mode of Production: The Case of Tanzania." In *Agricultural Development in Africa: Issues of Public Policy,* edited by Robert H. Bates and Michael F. Lofchie. New York: Praeger Publishers.
IBRD (International Bank for Reconstruction and Development)
1974. *Senegal: Tradition, Diversification, and Economic Development.*
1975. *Kenya: Into the Second Decade.*
1976. *Economic Memorandum on Sudan.*
1977. *Tanzania: Basic Economic Report,* Annex VI, Key Issues in Agriculture and Rural Development. Typescript.
1978a. *Ivory Coast: The Challenge of Success.*
1978b. "Nigeria: An Informal Survey." Mimeographed.
1979a. *World Development Report, 1979.*
1979b. *World Tables 1979.*
ICO (International Coffee Organization)
1978a. *Coffee in Kenya, 1977.*

1978b.   *Coffee in Tanzania, 1978.*

International Food Policy Research Institute
1977.   "Food Needs of Developing Countries: Projections of Production and Consumption to 1990." In *Research Report 3.* Washington, D.C.

IITA (International Institute for Tropical Agriculture)
1977.   *The National Accelerated Food Production Project: A New Dimension for Nigerian Development.*

ILO (International Labor Organization)
1975a.   *Growth, Employment and Equity: A Comprehensive Strategy for Sudan.* Volume II. Technical Paper No. 1. "Animal Husbandry, Migratory and Sedentary Populations."

1975b.   *Growth, Employment and Equity: A Comprehensive Strategy for Sudan.* Volume II. Technical Paper No. 2, "Irrigated Agriculture."

1975c.   *Growth, Employment and Equity: A Comprehensive Strategy for Sudan.* Volume II. Technical Paper No. 3, "Mechanized (rainfed) Agriculture."

1975d.   *Growth, Employment and Equity: A Comprehensive Strategy for Sudan.* Volume II. Technical Paper No. 17, "Incentives for Resource Allocation."

1975e.   *Growth, Employment and Equity: A Comprehensive Strategy for Sudan.* Volume II. Technical Paper No. 20, "Income Distribution."

1978.   *Employment, Income Distribution, Poverty Alleviation and Basic Needs in Kenya.* Typescript.

Jamal, Vali
1976.   "The Role of Cotton and Coffee in Uganda's Economic Development." Ph.D. Dissertation, Stanford University, Stanford, Calif.

Jeffries, Richard
1978.   *Class, Power and Ideology in Ghana: The Railwaymen of Sekondi.* Cambridge: Cambridge University Press.

Johnston, Bruce F.
1980.   "Agricultural Production Potentials and Small Farm Strategies in Sub-Saharan Africa." In *Agricultural Development in Africa: Issues of Public Policy,* edited by Robert H. Bates and Michael F. Lofchie. New York: Praeger Publishers.

Jones, William O.
1972.   *Marketing Staple Food Crops in Tropical Africa.* Ithaca and London: Cornell University Press.

1980.   "Agricultural Trade Within Tropical Africa: Historical Background."
        In *Agricultural Development in Africa: Issues of Public Policy*,
        edited by Robert H. Bates and Michael F. Lofchie. New York:
        Praeger Publishers.

Jorgenson, Dale W.
1969.   "The Role of Agriculture in Economic Development: Classical vs.
        Neoclassical Models of Growth." In *Subsistence Agriculture and
        Economic Development*, edited by Clifton R. Wharton. Chicago:
        Aldine Publishing Company.

Kaplinsky, Raphael
1978.   "Trends in the Distribution of Income in Kenya, 1966–1967."
        Mimeographed.

Kaneda, Hiromitsu, and Bruce F. Johnston
1961.   "Urban Food Expenditure Patterns in Tropical Africa." *Food Re-
        search Institute Studies* 2, No. 3:229–275.

Kenya
1966.   *Report of the Maize Commission of Inquiry, June 1966.*

1971.   *Report of the Working Party on Agricultural Inputs.*

1972.   *Report of the Select Committee on the Maize Industry.*

1977.   Central Bureau of Statistics. *Statistical Abstract, 1977.*

1970–77.   Coffee Board of Kenya. *Annual Report for the Year[s] Ended
        30th September 1969–76.*

KNFU (Kenya National Farmers' Union)
1971–78.   *Annual Reports 1969-70–1976-77.* Mimeographed.

Kenya Coffee Growers Association
1968.   "The Economic Situation of the Coffee Industry of Kenya Today."
        Typescript.

Kilby, Peter
1969.   *Industrialization in an Open Economy: Nigeria 1945–1966.* Cam-
        bridge: Cambridge University Press.

Killick, Tony
1978.   *Development Economics in Action.* New York: St. Martin's
        Press.

Kline, C. K., D. A. G. Green, Roy L. Donahue, and B. A. Stout
1969.   *Agricultural Mechanization in Equatorial Africa.* East Lansing:
        Michigan State University.

Knight, J. B.
1972.   "Rural Income Comparison and Migration in Ghana." *Bulletin,
        Oxford University Institute of Economics and Statistics* 34, 2:
        199–228.

Kotey, R. A., C. Okali, and B. E. Rourke
1974. *Economics of Cocoa Production and Marketing.* Legon, Ghana: Institute of Statistical, Social and Economic Research, University of Ghana.
Kraus, Jon
1976. "African Trade Unions: Progress or Poverty?" *African Studies Review* 19, 3:95–108.
Kriesel, Herbert C., Charles K. Laurent, Carl Halpern, and Henry E. Larzelere
1970. *Agricultural Marketing in Tanzania, Background Research and Policy Proposals.* East Lansing: Michigan State University.
Krueger, Anne O.
1974. "The Political Economy of the Rent-Seeking Society." *American Economic Review* 64, No. 3:291–303.
Langdon, Steven
1978. "The Multinational Corporation in the Kenya Political Economy." In *Readings on the Multinational Corporation in Kenya,* edited by Raphael Kaplinsky, pp. 134–200. Nairobi: Oxford University Press.
Leith, J. Clark
1974. *Foreign Trade Regimes and Economic Development: Ghana.* A Special Conference Series on Foreign Trade Regimes and Economic Development, Volume II. New York: National Bureau of Economic Research, Columbia University Press.
Leonard, David K.
1977. *Reaching the Peasant Farmer: Organization Theory and Practice in Kenya.* Chicago and London: University of Chicago Press.
Leubuscher, Charlotte
1956. *Bulk Buying From the Colonies.* London: Oxford University Press.
LeVine, Victor T.
1975. *Political Corruption: The Ghana Case.* Stanford, Calif.: Hoover Institution Press.
Lewis, W. Arthur
1963. "Economic Development with Unlimited Supplies of Labour." In *The Economics of Underdevelopment,* edited by A. N. Agarwala and S. P. Singh. New York: Oxford University Press.
Leys, Colin
1975. *Underdevelopment in Kenya.* London: Heinemann.
Libby, Ronald T.
1976. "External Co-optation of a Less Developed Country's Policy

Making: The Case of Ghana, 1969–1972." *World Politics* 29, 1: 67–89.

Lipton, Michael
1977. *Why Poor People Stay Poor: Urban Bias in World Development.* Cambridge, Mass.: Harvard University Press.

Lofchie, Michael F.
1975. "Political and Economic Origins of African Hunger." *The Journal of Modern African Studies* 13, No. 4:551–567.
1978. "Agrarian Crisis and Economic Liberalization in Tanzania." *Journal of Modern African Studies* 16, 3:451–475.

Macrae, David S.
1973. "Import Licensing in Kenya." Working Paper No. 90. Institute for Development Studies, University of Nairobi.

Maimbo, Fabian J. M., and James Fry
1971. "An Investigation into the Change in the Terms of Trade between the Rural and Urban Sectors of Zambia." *African Social Research* No. 12:95–110.

Maitha, J. K.
1969. "A Supply Function for Kenya Coffee." *Eastern Africa Economic Review* 1, No. 1:63–72.

Markovitz, Irving Leonard
1977. *Power and Class in Africa.* Englewood Cliffs, N.J.: Prentice-Hall.

McHenry, Dean E., Jr.
1979. *Tanzania's Ujamaa Villages.* Berkeley, Calif.: Institute of International Studies.

Migot-Adholla, S. E.
1979. "Rural Development Policy and Equality." In *Politics and Public Policy in Kenya and Tanzania*, edited by Joel D. Barkan with John J. Okumu, pp. 154–178. New York: Praeger Publishers.

Miracle, Marvin P., and Ann Seidman
1968. "State Farms in Ghana." Paper No. 43. Madison: University of Wisconsin Land Tenure Center.

Moore, Barrington, Jr.
1966. *Social Origins of Dictatorship and Democracy.* Boston: Beacon Press.

Moris, Jon R.
1974. "The Voter's View of the Election." In *Socialism and Participation: Tanzania's 1970 National Elections*, edited by The Election Study Committee, University of Dar es Salaam, pp. 312–364. Dar es Salaam: Tanzania Publishing House.

Morrison, Thomas K.
  1977. "The Political Economy of Export Instability in Developing
        Countries: The Case of Ghana." Paper presented at the Annual
        Meeting of the African Studies Association, Houston, Texas.
        Mimeographed.
Mwanasu, Bismarck U., and Cranford Pratt, eds.
  1979. *Towards Socialism in Tanzania.* Toronto: University of Toronto
        Press.
Nelson, Joan
  1979. *Access to Power: Politics and the Urban Poor in Developing Na-
        tions.* Princeton, N.J.: Princeton University Press.
Nicholson, Norman K., John D. Esseks, and Ali Akhtar Khan
  1979. "The Politics of Food Scarcities in Developing Countries." In
        Food, Politics, and Agricultural Development, edited by Ray-
        mond Hopkins, Donald J. Puchala, and Ross B. Talbot. Boulder,
        Col.: Westview Press.
Nigeria
  1962. *Report of the Coker Commission of Enquiry into the Affairs of
        Certain Statutory Corporations in Western Nigeria.*
  1974. *Public Service Review Commission: Main Report (Udoji Report).*
  1971. Federal Ministry of Information. *Second and Final Report of
        the Wages and Salaries Review Commission, 1970–71 (Adebo
        Report).*
  1975a. Federal Ministry of Information. *The Attack on Inflation: Gov-
        ernment Views on the First Report of the Anti-Inflation Task
        Force.*
  1975b. Federal Ministry of Information. *First Report of the Anti-Infla-
        tion Task Force.*
  1976. Federal Ministry of Information. *Report of the Special Panel on
        the Farms of Alhaji Andu Bako and Mr. S. O. Ogbemudia.*
Nigerian Economic Society
  1973. *Rural Development in Nigeria: Proceedings of the 1972 Annual
        Conference of the Nigerian Economics Society.* Ibadan: Nigerian
        Economics Society.
Njonjo, Apollo I.
  1977. "The Africanization of the 'White Highlands': A Study in Agrarian
        Class Struggles in Kenya, 1950–1974." Ph.D. Dissertation, Prince-
        ton University.
Northern Nigeria
  1967. *A White Paper on the Northern Nigeria Military Government's*

Policy for the Comprehensive Review of Past Operation and Methods of the Northern Nigeria Marketing Board. Kaduna: Government Printer.

Nyanteng, V. K.
1978. "Some Policies and Programs Related to Small Scale Farming in Ghana." Paper presented at a Policy Workshop on the Future Viability of Small Scale Farming. The Hague, Netherlands: Institute of Social Studies. Mimeographed.
1979. "Ghana: A Country Review Paper." Presented to the World Congress of Agrarian Reform and Rural Development. The Hague, Netherlands. Mimeographed.

Obben, James
1976. "A Study on the Costs of Processing Coconut at the Esiama Oil Mill and the Economic Viability of the Venture." Dissertation, Department of Agricultural Economics, University of Ghana.

O'Donnell, Guillermo
1978. "State and Alliances in Argentina, 1956–1976." *The Journal of Development Studies* 15, No. 1:4–33.
1979a. *Modernization and Bureaucratic Authoritarianism.* Berkeley, Calif.: Institute of International Studies.
1979b. "Tensions in the Bureaucratic-Authoritarian State and the Question of Democracy." In *The New Authoritarianism in Latin America,* edited by David Collier. Princeton, N.J.: Princeton University Press.

Ofosu, Samual Baffour
1972. "Case Study of a Farm Within the Workers Brigade." Dissertation, Department of Agricultural Economics, University of Ghana.

Okali, C., and E. Boreti-Doku
1977. "The Impact of the NAFPP Program in Imo State." Ibadan: International Institute for Tropical Africa.

Okoli, Angus
1978. "New Hopes for Agriculture." *Africa* No. 86:125.

Okoth-Ogendo, H. W. O.
1976. "African Land Tenure Reform." In *Agricultural Development in Kenya: An Economic Assessment,* edited by Judith Heyer, J. K. Maitha, and W. M. Senga, pp. 153–185. Nairobi: Oxford University Press.

Olatunbosun, Dupe
1975. *Nigeria's Neglected Rural Majority.* Ibadan: Oxford University

Press, for the Nigerian Institute of Social and Economic Research.

Oluwasanmi, H. A.
1966. *Agriculture and Nigerian Economic Development.* Ibadan: Oxford University Press.

Onitiri, H. M. A., and Dupe Olatunbosun
1974. *The Marketing Board System.* Ibadan: Nigerian Institute of Social and Economic Research.

Oyejide, T. Ademola
1975. *Tariff Policy and Industrialization in Nigeria.* Ibadan: Ibadan University Press.

Pearson, Scott R., Gerald C. Nelson, and J. Dirck Stryker
1976. "Incentives and Comparative Advantage in Ghanaian Industry and Agriculture." Paper for the West African Regional Project. Mimeographed.

Posner, Richard A.
1975. "The Social Costs of Monopoly and Regulation." *Journal of Political Economy* 83, No. 4:807–827.

Raikes, Philip
1978. "Rural Differentiation and Class Formation in Tanzania." *Journal of Peasant Studies* 5, No. 3:285–325.

Ranis, G. and J. C. H. Fei
1961. "A Theory of Economic Development." *American Economic Review* 51:533–565.

Roider, Werner
1971. *Farm Settlements for Socio-Economic Development: The Western Nigerian Case.* München: Weltforum Verlag.

Rothchild, Donald
1977. "Comparative Public Demand and Expectation Patterns: The Ghana Experience." Paper presented to the Annual Meeting of the American Political Science Association, Houston, Texas. Mimeographed.
1979. "Military Regime Performance: An Appraisal of the Ghana Experience, 1972–78." Paper prepared for delivery to the Annual Meeting of the American Political Science Association. Mimeographed.

Rweyemamu, Justin
1973. *Underdevelopment and Industrialization in Tanzania: A Study of Perverse Capitalist Industrial Development.* Nairobi: Oxford University Press.

Sabot, R. H.
1979. *Economic Development and Urban Migration.* Oxford: Clarendon
Press.
Sandbrook, Richard
1977. "The Political Potential of African Urban Workers." *Canadian
Journal of African Studies* 11, 3:411–433.
————, and Robin Cohen, eds.
1975. *The Development of an African Working Class: Studies in Class
Formation and Action.* London: Longman Group.
Saul, John S.
1979. *The State and Revolution in East Africa.* New York and London:
Monthly Review Press.
Schatz, Sayre P.
1970. *Economics, Politics and Administration in Government Lending:
The Regional Loans Boards of Nigeria.* Ibadan: Oxford University
Press, for the Nigerian Institute of Social and Economic Re-
search.
1977. *Nigerian Capitalism.* Berkeley and Los Angeles: University of
California Press.
Schultz, Theodore W.
1976. *Transforming Traditional Agriculture.* New York: Arno Press.
————, ed.
1978. *Distortion of Agricultural Incentives.* Bloomington, Ind.: Indiana
University Press.
Schumacher, Edward J.
1975. *Politics, Bureaucracy, and Rural Development in Senegal.* Berke-
ley and Los Angeles: University of California Press.
Scudder, Thayer
Forthcoming. "Policy Implications of Compulsory Relocation in River
Basin Development Projects." In *Projects for Rural Development:
The Human Dimension,* edited by Michael Cernea and Peter
Hammond. Baltimore, Md.: John Hopkins University Press.
1980. "River Basin Development and Local Initiative in African Savanna Environments." In *Human Ecology in Savanna Environments,* edited by D. R. Harris. London: Academic Press.
Sharpley, Jennifer
1976. "Intersectoral Capital Flows and Economic Development: Evidence from Kenya." Ph.D. Dissertation, Northwestern University.
1978. "Terms of Trade." Mimeographed, Nairobi.

164   Bibliography

Shivji, Isaa G.
1976.   *Class Struggles in Tanzania.* Dar es Salaam: Tanzania Publishing House.

Stiglitz, Joseph E., and Hirofumi Uzawa, eds.
1969.   *Readings in the Modern Theory of Economic Growth.* Cambridge, Mass. and London: The Massachusetts Institute of Technology Press.

Stryker, J. Dirck
1975.   "Ghana Agriculture." Paper for the West African Regional Project. Mimeographed.

Sudan, Democratic Republic of
1977.   Ministry of National Planning. *The Six Year Plan of Economic and Social Development 1977-78–1982-83.* Volume II.

Swainson, Nicola
1977a.   *Foreign Corporations and Economic Growth in Kenya.* Manuscript.
1977b.   "The Rise of a National Bourgeoisie in Kenya." *Review of African Political Economy* 8:21–38.
1978.   "State and Economy in Post-Colonial Kenya, 1963–1978." *Canadian Journal of African Studies* 12, No. 3:357–381.

Tanzania
1966.   *Report of the Special Committee of Enquiry into Co-operative Movement and Marketing Boards.*
1976.   Ministry of Agriculture. *Price Policy Recommendations for the 1977–1978 Agricultural Price Review.* Volume II. Mimeographed.
1977a.   Ministry of Agriculture. *Price Policy Recommendations for the 1978–1979 Agricultural Price Review.* Volume I, Summary and Price Proposals. Mimeographed.
1977b.   Ministry of Agriculture. *Price Policy Recommendations for the 1978–1979 Agricultural Price Review.* Volume II, Review of the New Crop Buying Arrangements. Mimeographed.
1977c.   Ministry of Agriculture. *Price Policy Recommendations for the 1978–1979 Agricultural Price Review.* Annex 1, Cereals. Mimeographed.
1977d.   Ministry of Agriculture. *Price Policy Recommendations for the 1978–1979 Agricultural Price Review.* Annex 9, Sisal. Mimeographed.
1977e.   Ministry of Agriculture. *Price Policy Recommendations for the 1978–1979 Agricultural Price Review.* Annex 10, Coffee. Mimeographed.

1974. University of Dar es Salaam. The Election Study Committee. *Socialism and Participation: Tanzania's 1970 National Elections.*

Temu, Peter E.

1975. "Marketing Board Pricing and Storage Policy with Particular Reference to Maize in Tanzania." Ph.D. Dissertation, Stanford University, Stanford, Calif.

Uganda

1929. *Report of the Commission of Enquiry into the Cotton Industry of Uganda.* Entebbe: Government Printer.

1938. *Report of the Uganda Cotton Commission.* Entebbe: Government Printer.

1948. *Report of the Cotton Industry Commission.* Entebbe: Government Printer.

1966. *Report of the Committee of Enquiry into the Cotton Industry.* Entebbe: Government Printer.

1967. *Report of the Committee of Enquiry into the Coffee Industry.* Entebbe: Government Printer.

United Nations

1978. *Statistical Yearbook 1977.*

USAID (United States Agency for International Development).

1975. *Development Assistance Program FY 1976–FY 1980, Ghana.* Volume 4, Annex D, Agricultural Sector.

1976. *Project Paper for Ghana, Managed Inputs and Delivery of Agricultural Services Programs for Small Farm Development.*

1978a. *Background Data on the Ivorian Rice Economy.*

1978b. *Rice Policy in Liberia.*

1978c. *Rice Policy in Mali.*

United States Government

1975. General Accounting Office. *Disincentives to Agricultural Production in Developing Countries.* Washington, D.C.

Usoro, Eno J.

1974. *The Nigerian Palm Oil Industry.* Ibadan: Ibadan University Press.

Walker, David, and Cyril Ehrlich

1959. "Stabilization and Development Policy in Uganda: An Appraisal." *Kyklos* 12:341–353.

Warren, Bill

1973. "Imperialism and Capitalist Industrialization." *New Left Review* 81:3–44.

Wells, Jerome C.

1974. *Agricultural Policy and Economic Growth in Nigeria, 1962–68.*

Ibadan: Oxford University Press, for the Nigerian Institute of Social and Economic Research.

Western Nigeria
1962. Ministry of Trade and Industry. *Report of the Commission of Enquiry into the Alleged Failure or Miscarriage of Plans to Effect a Revision of the Producer Price of Cocoa in January 1961.*

Western State, Nigeria
1969. *Report of the Commission of Enquiry into the Civil Disturbances which Occurred in Certain Parts of the Western State of Nigeria in the Month of December 1968.* Ibadan: Government Printer.

Whetham, Edith H.
1972. *Agricultural Marketing in Africa.* London: Oxford University Press.

Williams, Gavin
1976. "Nigeria: A Political Economy." In *Nigeria: Economy and Society,* edited by Gavin Williams, pp. 11–53. London: Rex Collings.

Young, Alastair
1973. *Industrial Diversification in Zambia.* New York: Praeger Publishers.

Zambia
1971. Central Statistical Office. *Census of Industrial Production, 1969.*

Index

27–28, 42–43. *See also* Bribes;
Smuggling
Cost of living, 31, 33–34, 35
Cotton, 26, 85; export of, 1, 11, 12,
137–38; mills, 25, 71; and the Gezira
scheme, 47–48 and *n*; prices for,
48*n*, 124
Cotton seed oil, 11
Coups d'etat: and economic unrest,
31–32
Courts: government control of, 106, 107
CPP. *See* Convention People's Party
Credit: subsidies on, 50, 52–53, 56–58,
59, 60*n*, 69, 91, 92, 94, 111. *See also*
Loans, agricultural; Loans, marketing
board
Credit agency, 111, 112
Crop authorities. *See* Marketing boards
Crops. *See* Cash crops; Export crops;
Food crops
Cross elasticity, 83–84
Currency: valuation of, 31, 35–36, 37,
101, 102, 131

Dairy products, 37
Depreciation allowances: and industrial
concentration, 69
Devaluation, 31, 102
Development agencies, 100
Development programs, 97, 113, 121,
123–24
Djin, A. K., 101
Doe, Samuel K., 105
Domestic competition. *See* Competition, domestic
Droughts, 1, 84

East Africa: and marketing boards, 13
Eggs, 46
Elasticities: of labor supply, 87
Electrical industry, 71
Elites: *vs.* peasants, 7, 43–44; and land,
60, 107; and markets, 100–03; and
rents, 103; and industry, 120; and ex-

port crops, 131–32; and imports,
131–32
Elites, political: and government policy,
6, 121; and workers, 33; and industry,
103–04, 119; and agricultural policy,
108, 126, 131–32. *See also*
Bureaucrats
Elites, urban: as food producers, 56,
58–59
Elites as farmers, 60–61, 128; *vs.* peasants, 7, 43–44, 113; and government
policy, 43–44, 45, 49, 60–61, 112–13,
120–21; and subsidies, 55, 56, 57–58,
75, 81, 126, 129; political power of,
60–61; as political opposition,
112–13. *See also* Farmers, large-scale
Ethiopia, 59
Exchange rate, 35–36
Export agriculture, 4, 11–29, 135; decrease in, 1–2; revenues from, 17,
18–19, 100, 128; and foreign exchange, 18, 134; taxes on, 123. *See also*
Cash crops; Export crops; Food crops
Export crops: and government marketing policy, 11–12, 25, 29, 40, 125–26,
136–45; and government pricing policy, 29, 109, 115, 123–24, 127,
136–45; compared to food crops, 40,
126; taxes on, 123; adverse incentives
for, 129; scenarios for government
policy concerning, 131; and elites,
131–32. *See also* Cash crops; Export
agriculture; Food crops
Export duties, 37
Exports: types of, 11, 21, 23, 24, 25–26,
37, 48*n*, 77; and marketing boards,
12, 134; and domestic prices, 36; banning of, 36, 37, 42, 126; and industrial
growth, 77; and foreign exchange, 102
Extension agencies, 49, 53, 55*n*. *See
also* Public services

Farmers: and agricultural exports, 1, 26,
136–45; and government policy, 2,

Designer: Sandy Drooker
Compositor: G & S Typesetters, Inc.
Printer: Murray Printing Company
Binder: Murray Printing Company
Text: Linotron 202 Caledonia
Display: Linotron 202 Bodoni